Charles Sprague Sargent

Forest Flora of Japan

Notes on the Forest Flora of Japan

Charles Sprague Sargent

Forest Flora of Japan
Notes on the Forest Flora of Japan

ISBN/EAN: 9783337163792

Printed in Europe, USA, Canada, Australia, Japan

Cover: Foto ©berggeist007 / pixelio.de

More available books at **www.hansebooks.com**

FOREST FLORA OF JAPAN

NOTES ON THE FOREST FLORA OF JAPAN

BY

CHARLES SPRAGUE SARGENT

DIRECTOR OF THE ARNOLD ARBORETUM OF HARVARD UNIVERSITY

BOSTON AND NEW YORK
HOUGHTON, MIFFLIN AND COMPANY
The Riverside Press, Cambridge
1894

THE following notes, gathered in the autumn of 1892, during a journey through Hondo and Yezo, were first printed, with the illustrations that accompany them, in "Garden and Forest." I am indebted to the publishers of that journal for the permission to reprint them in this form.

C. S. SARGENT.

ARNOLD ARBORETUM, *May, 1894.*

TABLE OF CONTENTS.

	Page
INTRODUCTION	1
THE MAGNOLIA FAMILY	8
THE TERNSTRŒMIA, LINDEN, AND RUE FAMILIES	17
THE HOLLY, EVONYMUS, AND BUCKTHORN FAMILIES	23
THE MAPLE FAMILY	28
THE SUMACHS AND THE PEA FAMILY	33
THE ROSE FAMILY	36
THE WITCH-HAZEL AND ARALIA FAMILIES	42
THE CORNELS, HONEYSUCKLES, AND PERSIMMONS, THE STYRAX FAMILY, THE ARBORESCENT MEMBERS OF THE HEATH FAMILY, THE ASHES, AND THEIR ALLIES	47
THE LAUREL, EUPHORBIA, AND NETTLE FAMILIES	54
THE WALNUTS, BIRCHES, ALDERS, AND HORNBEAMS	60
THE OAKS, CHESTNUTS, WILLOWS, AND POPLARS	67
THE CONIFERS	72
THE CONIFERS, II.	79
THE ECONOMIC ASPECTS OF THE JAPANESE FORESTS	85

LIST OF ILLUSTRATIONS.

1. Cercidiphyllum Japonicum.
2. Hemlock Forest of Lake Yumoto.
3. Magnolia Kobus.
4. Magnolia salicifolia.
5. Michelia compressa.
6. Cercidiphyllum Japonicum.
7. Cercidiphyllum Japonicum.
8. Tilia Miqueliana.
9. Acer Miyabei.
10. Acer Nikoense.
11. Gleditsia Japonica.
12. Prunus Maximowiczii.
13. Pyrus Miyabei.
14. Pyrus Tschonoskii.
15. Disanthus cercidifolia.
16. Acanthopanax ricinifolium.
17. Lindera obtusiloba.
18. Ulmus campestris.
19. Zelkova Keaki.
20. Alnus Japonica.
21. Carpinus Carpinus.
22. Ostrya Japonica.
23. Quercus dentata.
24. Avenue of Cryptomerias at Nikkō.
25. Hemlock Forest (Tsuga diversifolia), Lake Yumoto.
26. Larix Dahurica, var. Japonica.

FOREST FLORA OF JAPAN.

INTRODUCTION.

MANY years ago, in one of the most interesting papers [1] which has been written on the distribution of forests, Professor Asa Gray drew some comparisons between the forests of eastern North America and those of the Japan-Manchurian region of Asia. Here it was shown that, rich as eastern America is in tree species, Japan, and the regions to the north of it, in spite of their comparatively small area, are still richer. Professor Gray's Asiatic region included the four principal Japanese islands, eastern Manchuria, and the adjacent borders of China, while the contrasted American region embraced the territory east of the Mississippi River, but excluded the extreme southern point of Florida, inhabited by some sixty tropical trees which belong to the West Indian rather than to the true North American flora. In the Japan-Manchurian region he found 168 trees divided among sixty-six genera, and in eastern America 155 trees in sixty-six genera, the enumeration in both cases being confined "to timber-trees, or such as attain in the most favorable localities to a size which gives them a clear title to the arboreous rank." In the Japanese enumeration were included, however, a number of trees which are not indigenous to Japan, but which, as we now know, were long ago brought into the empire from China and Corea, like most of the plants cultivated by the Japanese. Early European travelers in Japan, like Thunberg and Siebold, who were unable to penetrate far into the interior, finding a number of plants common in cultivation, naturally believed them to be indigenous, and several Chinese plants were first described from individuals cultivated in Japanese gardens. Later writers [2] on the Japanese flora have generally followed the example of the early travelers, and included these plants in the flora of Japan. Indeed, it is only very recently that it has been possible to travel freely in all parts of the empire, and to study satisfactorily the character and distribution of its flora.

The list of Chinese and Corean trees cultivated in Japan, and usually enumerated in Floras of the empire, includes Magnolia conspicua, Magnolia parvifolia, Magnolia Watsoni, Sterculia platinifolia, Cedrela Sinensis, Zizyphus vulgaris, Kœlreuteria paniculata, Sapindus Mukorosi, Acer trifidum, Rhus vernicifera, Sophora Japonica,[3] Prunus Mume, Pyrus Sinensis, Cratægus cuneata, Eriobotrya Japonica, Liquidambar Formosana (Maximowiczii), Cornus officinalis,

---

[1] Forest Geography and Archæology, *Scientific Papers*, ii. 204.

[2] See Franchet & Savatier, *Enum. Pl. Jap.* — Forbes & Hemsley, *Jour. Linn. Soc.*, xxiii. and xxvi.

[3] Even Rein (*The Industries of Japan*), usually a most careful observer, states that Sophora Japonica is "scattered through the entire country, especially in the foliaceous forests of the north." He had evidently confounded Sophora with Maackia, a common and widely spread tree, especially in Yezo. Sophora, which is only seen occasionally in gardens, does not appear to be a particularly popular plant with the Japanese.

Diospyros Kaki, and probably Diospyros Lotus, Chionanthus retusa, Paulownia imperialis, Catalpa ovata, Lindera strychnifolia, Ulmus parvifolia, Thuja orientalis, Ginkgo biloba, Podocarpus Nageia, Podocarpus macrophylla, and Pinus Koraiensis. If these species,[1] twenty-nine in number, are deducted from Professor Gray's enumeration, there will remain 139 species in fifty-three genera, or a smaller number of both genera and species than he credited to eastern America. This, however, does not alter the fact that the Japanese region for its area is unsurpassed in the number of trees which inhabit its forests.

Indeed, the superiority of the forests of Japan in the number of their arborescent species over those of every other temperate region, eastern North America included, in proportion to their area, has certainly never been fully stated, as perhaps I shall be able to show, having made two years ago a somewhat extended journey through the northern and central islands, undertaken for the purpose of studying Japanese trees in their relations to those of North America. The case, perhaps, can best be stated by following Professor Gray's method, and making a new census of the inhabitants of the Japan-Manchurian forests and of those of eastern America, as these two regions extend through nearly the same degrees of latitude, and possess somewhat similar climates, although Japan has the advantage of a more equally distributed rainfall and a more equable climate, and offers a far more broken surface than eastern America, with mountains twice the height of any of the Appalachian peaks.

As the true Atlantic forest extends west to the eastern rim of the mid-continental plateau, the American region, for purposes of proper comparison, may be extended to the western limit of the Atlantic tree-growth, although this will add to the American side of the account a few genera and species of Texas, like Kœberlinia, Ungnadia, Parkinsonia, Prosopis, Acacia, Chilopsis, and Pithecolobium, which Professor Gray did not include in the enumeration from which his deductions were made. The south Florida species are again omitted, and those plants which grow up with a single stem will be considered trees. In eastern North America, that is in the whole region north of Mexico and east of the treeless plateau of the centre of the continent, but exclusive of south Florida, 225 species of trees, divided among 134 genera, are now known. The Japan-Manchurian region includes eastern Manchuria, the Kurile Islands, Saghalin, and the four great Japanese islands, but for our purpose does not include the Loochoo group, which, although it forms a part of the Japanese empire politically, is tropical and subtropical in the character of its vegetation, which, moreover, is still imperfectly understood. In this narrow eastern border of Asia there are now known 241 trees divided among ninety-nine genera. The extra Japanese portion of the region contributes but little to the enumeration. In Saghalin, Fr. Schmidt[2] found only three trees which do not inhabit Yezo, and in Manchuria, according to Maximowicz[3] and Schmidt,[4] there are only eighteen

---

[1] A number of shrubs, familiar in western gardens, and usually supposed to be Japanese from the fact that they were first known to Europeans in Japan, or were first sent from that country, are also Chinese or Corean, and in Japan are only found in gardens or in the neighborhood of habitations. Among them are Clematis patens, Magnolia stellata, Magnolia obovata, Berberis Japonica, Citrus Japonica, Prunus tomentosa, Prunus Japonica, Spiræa Thunbergii, Rhodotypos kerrioides, Cercis Chinensis, Enkianthus Japonicus, Forsythia suspensa, Olea fragrans, Tecoma grandiflora, Daphne Genkwa, Edgeworthia papyrifera, and Wikstrœmia Japonica. Nandina domestica, the most universally cultivated ornamental plant in Japan, is probably not a Japanese plant, although Rein states that it grows wild in Shikoku.

[2] *Reisen in Amurland.*

[3] *Prim. Fl. Amur.*

[4] *Reisen in Amurland.*

INTRODUCTION. 3

trees which do not also occur in Saghalin or in the northern Japanese islands. In the four islands of Yezo, Hondo, Shikoku, and Kyûshû, therefore, we now find 220 trees divided among ninety-nine genera, or only five less than occur in the immense territory which extends from Labrador to the Rio Grande and from the shores of the Atlantic to the eastern base of the Rocky Mountains. Neither Cycas revoluta nor Trachycarpus (Chamærops) excelsa is included in the Japanese list, as the best observers appear to agree in thinking that these two familiar plants are not indigenous to Japan proper. I have omitted, moreover, a few doubtful species from the Japan enumeration, like Fagus Japonica, Maximowicz, and Abies umbellata, Mayr, of which I could learn nothing in Japan, so that it is more probable that the number of Japanese trees will be increased than that any addition will be made to the silva of eastern America.

The proportion of trees to the whole flora of Japan is remarkable, being about 1 to 10.14, the number of indigenous flowering plants and vascicular cryptogams being not very far from 2,500 species. Still more remarkable is the large proportion of woody plants to the whole flora. In Japan proper there are certainly not less than 325 species of shrubs, or 550 woody plants in all, or one woody plant in every 4.55 of the whole flora, — a much larger percentage than occurs in any part of North America.

The aggregation of arborescent species in Japan is, however, the most striking feature in the silva of that country. This is most noticeable in Yezo, where probably more species of trees are growing naturally in a small area than in any other one place outside the tropics, with the exception of the lower basin of the Ohio River, where, on a few acres in southern Indiana, Professor Robert Ridgway has counted no less than seventy-five arborescent species in thirty-six genera.[1] Near Sapporo, the capital of the island, in ascending a hill which rises only 500 feet above the level of the ocean, I noticed the following trees: Magnolia hypoleuca, Magnolia Kobus, Cercidiphyllum Japonicum, Tilia cordata, Tilia Miqueliana, Phellodendron Amurense, Picrasma ailanthoides, Evonymus Europæus, var. Hamiltonianus, Acer pictum, Acer Japonicum, Acer palmatum, Rhus semialata, Rhus trichocarpa, Maackia Amurensis, Prunus Pseudo-Cerasus, Prunus Ssiori, Pyrus aucuparia, Pyrus Toringo, Pyrus Miyabei, Hydrangea paniculata, Aralia spinosa, var. canescens, Acanthopanax ricinifolium, Acanthopanax sciadophylloides, Cornus macrophylla, Syringa Japonica, Fraxinus Mandshurica, Fraxinus longicuspis, Clerodendron trichotomum, Ulmus campestris, Ulmus scabra, var. laciniata, Morus alba, Juglans Sieboldiana, Betula alba, Betula alba, var. Tauschii, Betula alba, var. verrucosa, Betula Ermani, Betula Maximowicziana, Alnus incana, Carpinus cordata, Ostrya Japonica, Quercus crispula, Quercus gosseserrata, Castanea vulgaris, Populus tremula, Picea Ajanensis, Abies Sachalinensis, — forty-six species and varieties. Within five miles of this hill also grow Acer spicatum, var. Kurunduense, Acer Tataricum, var. Ginnala, Styrax Obassia, Aphananthe aspera, Quercus dentata, Quercus glandulifera, Alnus Japonica, Salix subfragilis, Salix Caprea, Salix stipularis, Salix acutifolia, Salix viminalis, and Populus suaveolans, — in all sixty-two species and varieties, or more than a quarter of all the trees of the empire, which are crowded into an area a few miles square, in the latitude of northern New England, in which, north of Cape Cod, there are only about the same number of trees.

[1] See *Proc. U. S. Nat. Mus.* 1882, 52. — *Garden and Forest*, vi. 148.

A further examination of the trees of the two countries shows that, although the Japan-Manchurian region possesses more arborescent species than eastern America, the silva of the latter is much richer in genera, — one hundred and thirty-four to ninety-nine in Japan-Manchuria. Forty-four genera have arborescent species in the two regions; forty-five genera with Japanese representatives have none in the flora of eastern America, and thirty-eight genera represented in the American flora do not appear in that of Japan. A few genera, five in eastern America and seven in Japan, are represented by trees in one region and by shrubs only in the other. Of endemic arborescent genera the silva of eastern America contains Asimina, Kœberlinia, Cliftonia, Ungnadia, Robinia, Cladrastis, Pinckneya, Oxydendrum, Mohrodendron, Sassafras, Planera, Toxylon, Leitneria, Hicoria, and Taxodium, fifteen, while in Japan there are only five, — Cercidiphyllum,[1] Trochodendron, Platycarya, Cryptomeria, and Sciadopitys.

Such a comparison between the silvas of eastern America and Japan is interesting as showing the great number of arborescent species inhabiting four small islands. The significant comparison, however, if it can ever be made, will be between eastern America, as here limited, and all of eastern Asia from the northern limits of tree-growth to the tropics, and from the eastern rim of the Thibetan plateau to the eastern coast of Japan. This would include Corea, practically an unexplored country botanically, especially the northern portions, and all the mountain ranges of western China, a region which, if it is to be judged from the collections made there in recent years, is far richer in trees than Japan itself. It is impossible to discuss with precision or with much satisfaction the distribution of the ligneous plants of the north temperate zone until more is known of western China and of Corea, where may be sought the home of many plants now spread through eastern China and Japan, and where alone outside the tropics the enterprising and industrious collector may now hope to be rewarded with new forms of ligneous vegetation.

Travelers in Japan have often insisted on the resemblance between that country and eastern America in the general features of vegetation. But with the exception of Yezo, which is still mostly uninhabited and in a state of nature, and those portions of the other islands which are over 5,000 feet above the level of the ocean, it is difficult to form a sufficiently accurate idea of the general appearance of the original forest-covering of Japan to be able to compare the aspects of its vegetation with those of any other country, for every foot of the lowlands and the mountain valleys of the three southern islands has been cultivated for centuries. And the foothills and low mountains which were once clothed with forests, and could be again, are now covered with coarse herbage, principally Eulalia, and are destitute of trees, except such as have sprung up in sheltered ravines, and have succeeded in escaping the fires which are set every year to burn off the dry grasses. Remoteness, bad roads, and the impossibility of bringing down their timber into the valleys have saved the mountain forests of Japan, which may still be seen, especially between 5,000 and 8,000 feet above the level of the sea, in their natural condition. But these elevated forests are composed of comparatively few species, and if it were not for the plantations of Conifers, which the Japanese for at least twelve centuries, it is said, have been making to supply their workers in wood with material, and for

---

[1] Since this was written I have received through M. M. L. de Vilmorin of Paris seeds of a Cercidiphyllum gathered in the extreme western part of China.

the trees preserved or planted in the temple grounds in the neighborhood of towns, it would be impossible to obtain any idea at all of many of the Japanese trees. But, fortunately, for nearly two thousand years the priests of Buddha have planted and replanted trees about their temples, which are often surrounded by what now appear to be natural woods, as no tree is ever cut and no attempt is made to clear up the undergrowth. These groves are sometimes of considerable extent, and contain noble trees, Japanese and Chinese, which give some idea of what the inhabitants of the forests of Japan were before the land was cleared for agriculture.

The floras of Japan and eastern America have, it is true, some curious features in common, and the presence in the two regions of certain types not found elsewhere show their relationship. But these plants are usually small, and are rare or grow only on the high mountains. Diphylleia, Buckleya, Epigæa, and Shortia show the common origin of the two floras; but these are rare plants in Japan, as they are in America, with the exception of Epigæa, and probably not one traveler in ten thousand has ever seen them, while the chief elements of the forest flora of northern Japan, the only part of the empire where, as has already been said, comparison is possible, — those which all travelers notice, — do not recall America so much, perhaps, as they do Siberia and Europe.

The broad-leaved Black Oaks, which form the most distinct and conspicuous feature in all the forests of eastern America, are entirely absent from Japan, and the deciduous-leaved White Oaks, which, in Japan, constitute a large part of the forest-growth of the north, are of the European and not of the American type, with the exception of Quercus dentata, which has no related species in America. The Chestnut Oaks, which are common and conspicuous, both in the northern and southern parts of eastern America, do not occur in Japan, and the Evergreen Oaks, which abound in the southern part of that empire, where they are more common than any other group of trees, are Asiatic and not American in their relationships.

Many of our most familiar American trees are absent from the forests of Japan. The Tulip-tree, the Pawpaw or Asimina, the Ptelea or Hop-tree, the Loblolly Bay or Gordonia, the Cyrilla and the Cliftonia, the Plum-trees, which abound here in many forms, the Texas Buckeye (Ungnadia), the Mesquite, the Locusts, the Cladrastis or Virgilia, the Kentucky Coffee-tree or Gymnocladus, the Liquidambar, the Tupelos, the Sourwood or Oxydendrum, the Osage Orange, the Kalmia, the Sassafras, the Persea or Red Bay, the Planera or Water Elm, the Plane-tree, the Black Walnut, the Hickories, and the deciduous Cypress — all common and conspicuous in our forests — are not found in Japan. Cratægus, with a dozen species, is one of the features of the forest flora of eastern America, while in Japan the genus is represented by a single species, confined to the northern part of the empire, and nowhere very common. The Japanese Maples, with the exception of Acer pictum, which is not unlike our Sugar Maple, have no close resemblance or relationship with the eastern American species; the Beech and the Chestnut are European, and not American; the Birches, with one exception, are of the Old World type, as are the Lindens, Ashes, Willows, the Celtis, the Alders, Poplars, and Larches.[1]

[1] Of the arborescent genera of Japan, thirty are represented in Europe, and all of these, with the exception of Buxus, are also found in eastern America.

On the other hand, the Japanese will not find in our forests Euptelea and Cercidiphyllum of the Magnolia family, Trochodendron, Idesia, the arborescent Ternstrœmiaceæ (Ternstrœmia, Cleyera, Eurya, and Camellia), Phellodendron and Hovenia, Euscaphis, Maackia and Albizzia, Distylium, Acanthopanax, Syringa, many arborescent Lauraceæ (Cinnamomum, Machilus and Actinodaphne), which, next to the Evergreen Oaks, are the most distinguishing features of the forest flora of southern Japan. Nor will they find the beautiful arborescent Linderas which abound in Japan, while in America the genus is only represented by two unimportant shrubs, the arborescent Euphorbiaceæ, like Buxus, Daphniphyllum, Aleurites, Mallotus, and Excœcaria, or Zelkova, Aphananthe, Broussonetia, and Debregeasia, or find anything to remind them of Pterocarya and Platycarya, Cryptomeria, Cephalotaxus, and Sciadopitys.

The forests of the two regions possess in common Magnolia and Æsculus, which are more abundant in species and individuals in America than in Japan. The Rhuses or Sumachs are very similar in the two regions, and so are the Witch Hazel and the arborescent Aralia. Cornus macrophylla of Japan is only an enlarged Cornus alternifolia of eastern America, and the so-called Flowering Dogwoods of the two countries are not unlike. The Japanese Walnut is very like the American Butternut, while, rather curiously, the Japanese Thuya and the two Chamæcyparis, the Piceas and Abies, resemble species of Pacific North America, a region whose flora has little affinity with that of eastern Asia. Tumion is common to the two regions; in eastern America it is one of the most local of all our trees, while in Japan it is abundant in the mountainous regions of the central and southern parts of the empire.

Apart from the characters which distinguish related genera and species of Japanese trees from their American congeners there are many aspects of vegetation which make the two countries unlike. The number of broad-leaved evergreen trees is much greater in southern Japan than in the southern United States, there being fifty species of these trees in the former, and only twenty in eastern America (exclusive always of southern Florida), and the general aspect of the groves and woods at the sea-level, even in the latitude of Tōkyō, is of broad-leaved evergreens. The number of evergreen shrubs in proportion to the entire flora is much greater in Japan, too, than it is in America, and plants of this character grow much farther north in the former than in the latter country. The small number of species of Pinus in Japan, and their scarcity at the north, is in striking contrast to the number and distribution of the species of this genus in eastern America, where there are thirteen species to only five in Japan, including one shrub. In Japan the Hemlock forms continuous and almost unbroken forests of great extent on the mountain-slopes, which are over 5,000 feet above the sea, while in eastern America this tree is rarely found except scattered in small groves or as single individuals through the deciduous-leaved forests. On the other hand, Picea and Abies, which in America form immense forests almost to the exclusion of other species, grow, wherever I have seen them in Japan, singly, or, in the case of Abies, in small groves on the lower border of the Hemlock forests or mingled with deciduous-leaved trees. Picea Ajanensis is said, however, to form extensive forests in some parts of western Yezo, and Professor Miyabe informs me that in the extreme northern part of that island there are fine continuous forests of Abies Sachalinensis. In northern Japan and on the high mountains of the central

PLATE II.

HEMLOCK FOREST OF LAKE YUMOTO.

islands, Birches are more abundant than they are in our northern forests; and the river banks at the north, like those of northern Europe and Siberia, are lined with arborescent Willows and Alders, which are rare in eastern America, where these genera are usually represented by shrubs.

The illustration on the opposite page (Plate ii.) gives some idea of the general appearance of the great coniferous forests which cover the highlands of central Japan. In the foreground, Lake Yumoto, famous for its thermal springs, nestles, 5,000 feet above the sea, among the Nikkō Mountains. The forests which rise from the shores of the lake are principally composed of Hemlock (Tsuga diversifolia), among which are Birch (Betula Ermani), Abies and Picea, Pterocarya, Cercidiphyllum, and the Mountain Ash. In the dense shade by the shores of the lake grow dwarf forms of the Indian Azalea, Elliottia paniculata, our Canadian Bunch Berry (Cornus Canadensis), great masses of Rhododendron Metternichii, which in these forests replaces Rhododendron Catawbiense of the Appalachian Mountains, the dwarf Ilex rugosa, Clethra canescens, here at the upper limits of its distribution, Panax horrida, and the dwarf Blueberries which inhabit mountain-slopes in all northern countries, as well as the ubiquitous Bamboos.

The undergrowth which covers the ground beneath the forests in the two regions is so unlike that it must at once attract the attention of the most careless observer. In all the Appalachian region of North America this is composed of a great number of shrubs, chiefly of various species of Vaccinium and Gaylusaccia, of Epigæa, wild Roses, Kalmias, dwarf Pyrus and Lycopodiums; in Japan the forest-floor is covered, even high on the mountains, and in the extreme north, with a continuous, almost impenetrable, mass of dwarf Bamboos of several species, which makes traveling in the woods, except over long-beaten paths and up the beds of streams, practically impossible. These Bamboos, which vary in height from three to six feet in different parts of the country, make the forest-floor monotonous and uninteresting, and prevent the growth of nearly all other under-shrubs, except the most vigorous species. Shrubs, therefore, are mostly driven to the borders of roads and other open places, or to the banks of streams and lakes, where they can obtain sufficient light to enable them to rise above the Bamboos; and it is the abundance of the Bamboo, no doubt, which has developed the climbing habit of many Japanese plants, which are obliged to ascend the trees in search of sun and light, for the Japanese forest is filled with climbing shrubs, which flourish with tropical luxuriance.

The wild Grape grows in the damp forests of Yezo with a vigor and to a size which the American species do not often attain, even in the semitropical climate of the southern Mississippi valley. Actinidia arguta climbs into the tops of the tallest trees, and nothing is so un-American or so attracts the attention of the American traveler in Japan as the trunks of trees clothed to the height of sixty or eighty feet with splendid masses of the climbing Hydrangeas (H. petiolaris and Schizophragma), or with the lustrous evergreen foliage of the climbing Evonymus. Wistaria is represented, it is true, in eastern America, but here it is not common or one of the chief features of vegetation as it is in Japan; and the Ivy, a southern plant only in Japan, and nowhere very common, helps to remind the traveler that he is in the Old and not in the New World.

THE MAGNOLIA FAMILY.

The general character of the composition of the Japanese forests having been briefly traced, I shall now say something of the most important Japanese trees; and, as their botanical characters are already pretty well understood and their economic properties are only of secondary interest to the general reader, these remarks will relate principally to their quality from a horticultural point of view. A comparison with allied eastern American species will perhaps be useful; it will, at any rate, show that, while Japan is extremely rich in the number of its tree species, the claim that has been made, that the forests of eastern America contain the noblest deciduous trees of all temperate regions, can, so far as Japan is concerned, be substantiated, for, with few exceptions, the deciduous trees of eastern America surpass their Asiatic relatives in size and beauty.

In the Magnolia family Japan possesses five genera, while in the United States there are only four. In Japan arborescent Magnoliaceæ reach the most northern limit attained in any country by these plants, and one of the most interesting features of the Japanese flora is the presence in Yezo of two large trees of this tropical and semitropical family as far north, at least, as the forty-fourth degree, while the representative of a third genus, Schizandra, is found still farther north on the Manchurian mainland. In eastern America two species of Magnolia reach nearly as high latitudes as this genus does in Japan, but in the United States Magnolia is really southern, and has only succeeded in obtaining a precarious foothold at the north, while in Yezo it is a most important element and a conspicuous feature of the forest vegetation.

Of true Magnolias three species grow naturally in Japan; two of these belong to the section of the genus which produces its flowers before the leaves appear, and which has no representative in the flora of America; the third, Magnolia hypoleuca, bears some resemblance to our Magnolia tripetala. This tree is seen at its best in the damp rich forests which cover the low rolling hills of Yezo, where it sometimes rises to the height of a hundred feet and forms trunks two feet in diameter; on the other Japanese islands it is confined to the mountain forests, and apparently does not descend below 2,000 feet above the sea; and it is only in Yezo and on the high mountains in the extreme northern part of the main island that I saw it of large size. In central Japan it rarely appears more than twenty or thirty feet high, although this can perhaps be accounted for by the fact that all trees in the accessible parts of the Japanese forests are cut as soon as they are large enough to be used for timber. Magnolia hypoleuca must be considered a northern species, requiring a cold winter climate for its best development, and it probably will not thrive in regions where the ground is not covered with snow during several months of every year.

Magnolia hypoléuca is one of the largest and most beautiful of the deciduous-leaved Magnolias; in the early autumn, when the cones of fruit, which exceed those of any of our species in size and are sometimes eight inches long, and brilliant scarlet in color, stand out on

PLATE III.

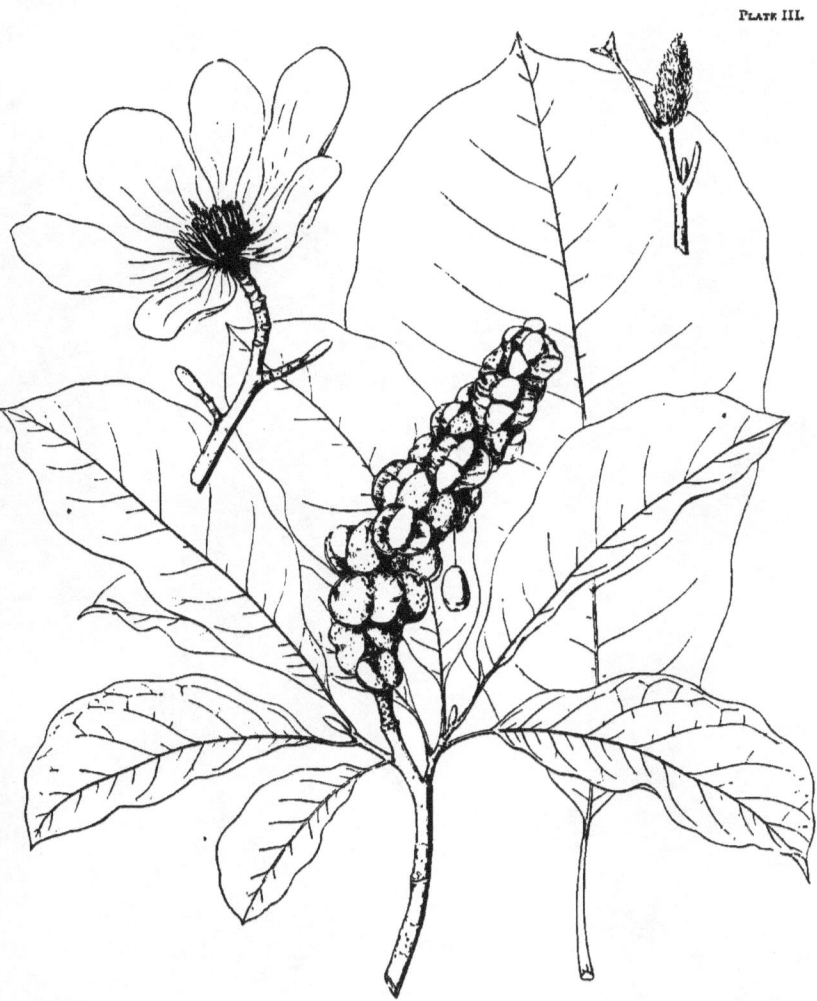

MAGNOLIA KOBUS, D. C.

the branches, it is the most striking feature of the forests of Hokkaido, which in variety and interest are not surpassed by those of any other part of the world. Like Magnolia tripetala, it is a tree of open habit, with long spreading irregularly contorted branches covered, as well as the trunk, with pale smooth bark. The leaves, however, are not as much crowded together at the ends of the flowering branches as they are in the American species, but are placed rather remotely on the branchlets; they are twelve or fourteen inches long and seven or eight inches broad, and on young vigorous trees are sometimes twice this size. On the upper surface they are light bright green, and pale steel blue or sometimes almost silvery on the lower surface, so that when raised by the wind they give the tree a light and cheerful appearance. The flowers are six or seven inches across when expanded, with creamy white petals and brilliant scarlet filaments; they appear in May and June, after the leaves are nearly full grown, and are very fragrant.

Magnolia hypoleuca is still rare in gardens, although it was sent by Mr. Thomas Hogg to the United States as early as 1865, and has flowered in the neighborhood of New York and Boston; it is probable that it will thrive in any part of the northern United States, although, like other Hokkaido trees, it may suffer from summer and autumn droughts, which are unknown in Japan, where the rainfall during August and September is regular and abundant. As an ornamental tree Magnolia hypoleuca is superior to Magnolia tripetala in the fragrance of its flowers and in the coloring of its leaves; it is less desirable than Magnolia macrophylla, which surpasses its Japanese relative in form and in the size and beauty of its flowers and leaves, the largest produced on any plant of the Magnolia family, and larger than those of any other North American tree. As a timber-tree Magnolia hypoleuca is valuable. The wood, like that of all the Magnolias, is straight-grained, soft, light-colored, and easily seasoned and worked. It is esteemed and much used in Japan for all sorts of objects that are to be covered with lacquer, especially sword-sheaths, which are usually made from it; in Hokkaido it is employed for the interior finish of houses, and for boxes and cabinets, although harder woods are generally preferred for such purposes.

In the forests of Hokkaido a second species, Magnolia Kobus,[1] occurs. This tree sometimes grows in the neighborhood of Sapporo to the height of seventy or eighty feet, and develops a tall straight trunk nearly two feet in diameter, covered with rather dark, slightly furrowed bark. The branches are short and slender, and form a narrow pyramidal head, which only becomes round-topped when the tree has attained its full size. The branchlets are more slender than those of most species of Magnolia, and are covered with dark reddish brown bark. The flowers (Plate iii.) appear near Sapporo in the middle of May, before the leaves, from acute buds an inch long, half an inch broad, and protected by long thickly matted pale hairs. They are from four to five inches across when fully expanded, with small acute caducous sepals and narrow obovate thin creamy white petals. The stamens, with short broad filaments, are much shorter than the narrow acute cone of pistils. The leaves are obovate,

---

[1] Magnolia Kobus, De Candolle, *Syst.* i. 456. — Miquel, *Prol. Fl. Jap.* 146. — Maximowicz, *Mél. Biol.* viii. 507. — Franchet & Savatier, *Enum. Pl. Jap.* i. 16.

Magnolia tomentosa, Thunberg, *Trans. Linn. Soc.* ii. 336 (in part).
Magnolia glauca, var. α, Thunberg, *Fl. Jap.* 236.
Kobus, Kaempfer, *Icon. Select.* t. 42.

gradually narrowed below, and abruptly contracted at the apex into short broad points; they are pubescent on the lower surface at first, especially on the stout midribs and primary veins, but at maturity are glabrous, or nearly so, and are bluish green, and rather lighter colored on the lower than on the upper surface; they are six or seven inches long, three or four inches broad, rather conspicuously reticulate-veined, and are borne on stout petioles half an inch to an inch and a half in length. The fruit is slender, four or five inches long, and is often contorted or curved from the abortion of some of the seeds; it is dark brown, the carpels being conspicuously marked with pale dots.

Magnolia Kobus is exceedingly common in the forests which clothe the hills in the neighborhood of Sapporo, where it grows to a larger size than in any part of Japan which I visited; near the shores of Volcano Bay it occurs in low swampy ground and in the neighborhood of streams, in situations very similar to those selected by Magnolia glauca in the United States. On the main island Magnolia Kobus is much less common than it is in Hokkaido, and I only met with it occasionally on the Hakone and Nikkō Mountains at considerable elevations above the sea. This handsome tree was introduced into the United States by Mr. Thomas Hogg, and was distributed from the Parsons' Nurseries as Magnolia Thurberi, under the belief that it was an undescribed species. In cultivation it does not flower freely in the young state, although trees in Pennsylvania, and in the Arnold Arboretum, where it was raised from seed sent from Sapporo fifteen years ago, have produced a few flowers. In New England Magnolia Kobus is the hardiest, most vigorous, and most rapid growing of all Magnolias.

I spent the 2d and 3d of October in company with Mr. James Herbert Veitch and Mr. Tokubuchi, an accomplished Japanese botanist, on Mount Hakkoda, an extinct volcano 6,000 feet high, which rises southeast and a few miles distant from Aomori, the most northern city of the main island of Japan. Botanically this was one of the most interesting excursions I made in Japan, and we were able to gather the seeds of a number of plants that we did not meet with elsewhere. On this mountain, in the very spot, perhaps, where Maries discovered this fine tree, we found Abies Mariesii covered with its large purple cones; and on the upper slopes saw the dwarf Pinus pumila, forming almost impenetrable thickets five or six feet high and many acres in area, and numerous alpine shrubs like Andromeda nana, Gaultheria pyroloides, Epigæa Asiatica, Phyllodoce taxifolia, and Geum dryadoides. On this mountain, too, we established the most northern recorded station in Asia of the Hemlock (Tsuga diversifolia); and near the base Ilex crenata, Ilex Sugeroki, a handsome evergreen species with bright red fruit, the dwarf Ilex integra, var. leucoclada, and Daphniphyllum humile were very common; and here we were fortunate in finding good fruit and ripe seeds of Magnolia salicifolia.

On Mount Hakkoda Magnolia salicifolia (see Plate iv.) is a common plant between 2,000 and 3,000 feet above the sea-level. As it appears there it is a slender tree fifteen or twenty feet high, with stems three or four inches thick, covered with pale smooth bark, and sometimes solitary, or more commonly in clusters of three or four. The branchlets are slender, and light green at first, like those of Magnolia glauca, later growing darker, and in their third year becoming dark reddish brown. The leaves are ovate, acute, gradually narrowed or rarely rounded at the base, contracted into long slender points and sometimes slightly falcate

MAGNOLIA SALICIFOLIA, Maxm.

toward the apex; they are thin, light green on the upper and silvery white on the lower surface, quite glabrous at maturity, five or six inches long, an inch and a half to two inches broad, and are borne on slender petioles half an inch in length. When bruised they are more fragrant than those of any species of Magnolia with which I am acquainted, exhaling the delicate odor of anise-seed. The winter flower-bud is two thirds of an inch long, rather obtuse, and protected by a thick coat of yellow-white hairs. The flowers of this tree are not known to botanists, but from the shape and character of the winter-buds they are probably of good size and produced in early spring before the appearance of the leaves. The fruit is slender, flesh-color, an inch and a half to two inches long, and half an inch broad.

Magnolia salicifolia[1] grows on Mount Hakkoda in low wet situations, generally near streams, and is evidently a moisture-loving plant. In November I found a single small plant of this species near the town of Fukushima, on the hills which rise above the valley of the Kisogawa, not far from the base of Mount Ontake, in central Japan. Magnolia salicifolia is new to cultivation, and we were fortunate in obtaining a good supply of seeds, by means of which, it is to be hoped, this interesting tree will soon appear in gardens.

Among the Magnolias of Japan there is no evergreen species which resembles the great evergreen Magnolia of our southern states, or at all equals it in the beauty of flowers and foliage; and the nearest approach to an evergreen Magnolia in the empire of the Mikado is the representative of a closely allied genus, Michelia, differing from Magnolia in the position of the flowers, which, instead of being terminal on the branches, are, except in the case of one Indian species, axillary; and in the number of ovules and seeds, of which there are two in each carpel of Magnolia and more than two in each carpel of Michelia. A dozen species are known, inhabitants of southern and southwestern Asia, including the islands of the Indian Archipelago, and, with the exception of the Chinese Michelia fuscata (or as it is habitually called in our southern states, Magnolia), which is cultivated for its exceedingly fragrant small flowers in all warm temperate countries, the genus is not seen in American or European gardens.

The Japanese species, Michelia compressa[2] (see Plate v.), as it appears in the Botanic Garden of the University of Tōkyō, is a tree thirty to forty feet in height, with a trunk twelve to eighteen inches in diameter, covered with smooth dark bark, and rather slender branches which form a compact handsome round-topped head. The winter-buds and the branchlets during their first year are clothed with soft ferruginous or pale hairs; in their second season the branchlets are slender, light or dark brown, marked, especially near their extremities, with large pale lenticels, and conspicuous from the raised nearly circular leaf-scars. The leaves are oblong or narrowly obovate, gradually contracted into long slender petioles, and are rounded or short-pointed at the apex, entire, coriaceous, conspicuously reticulate-veined, and dark green and lustrous on the upper, and pale and dull on the lower surface; they are three or four inches long, an inch to an inch and a half broad, with petioles an

---

[1] Magnolia salicifolia, Maximowicz, *Mél. Biol.* viii. 509. — Franchet & Savatier, *Enum. Pl. Jap.* i. 16.
Bürgeria (?) salicifolia, Siebold & Zuccarini, *Fl. Jap. Fam. Nat.* i. 187. — Miquel, *Prol. Fl. Jap.* 144.

[2] Magnolia (Michelia) compressa, Maximowicz, *Mél. Biol.* viii. 506. — Franchet & Savatier, *Enum. Pl. Jap.* i. 15.

inch in length, and fall, when a year old, after the appearance of the new shoots. The flowers, which are very fragrant, are from an inch to an inch and a quarter across when expanded, with pale yellow narrow obovate sepals and petals, nearly sessile anthers, and a stipitate head of pistils, the ovaries, according to Maximowicz, each containing five or six ovules. The cone of fruit is two inches long, and is raised on a stalk half an inch or more in length; the rusty brown thick-walled carpels, which are marked with large pale circular dots, usually containing three seeds; these are broadly ovate and much flattened by mutual pressure.

Michelia compressa, which is the most boreal species of its genus, was discovered near Nagasaki by Maximowicz, who saw a single tree. Oldham, an English botanist, collected it near the base of Fuji-san, doubtless from a cultivated tree, and it is said to be found in several places in the extreme southern part of the empire, although I have never seen it except in the Botanic Garden of Tōkyō, where there are several large trees, from which were collected the specimens figured in our illustration. It is not improbable that Michelia is at the northern limit of its range in Japan and that it will be found to be more at home on the Loochoo Islands or on Formosa when the interesting flora of these islands is carefully explored. The fact that Michelia compressa flourishes in Tōkyō, where our southern evergreen Magnolia hardly survives, indicates that it may perhaps be grown as far north as Washington and, possibly, Philadelphia, in southern England and Ireland, and on the west coast of France, that is, in regions where no other species of this genus can exist in the open ground, and where broad-leaved evergreen trees are rare and much desired.

The Japanese Illicium is a beautiful and interesting plant. It is the representative of a genus with two species in our southern states and half a dozen others in India and southern China, one of which, Illicium verum, supplies the star anise of the pharmacists. Illicium anisatum, or as it should, perhaps, be called, Illicium religiosum, is a beautiful small evergreen tree, fifteen or twenty feet high, with brilliant persistent leaves and small fragrant yellow flowers. It is one of the sacred plants of Japan. Siebold considered it a native of China or Corea and an introduction by Buddhist priests into Japan. This may be the correct view, although Japanese botanists now believe it to be a native of the southern part of the empire, where Rein found it, as he supposed, growing wild. Sacred to Buddha, it is always planted in the neighborhood of his temples, and is common in private gardens as far north as the thirty-fifth degree of latitude. The branches of this tree, especially when it is in flower, are used to decorate the altars in the temples, or in cemeteries serve to mark the respect of the living for the dead. From the powdered bark, mixed with resin, are prepared the "smoke candles" with which incense is made in the temples, and with which the "moxa" is burned on the human body as a sovereign cure for many of its ills.

The remaining Japanese plants of the Magnolia family, Kadsura and Schizandra, are woody climbers. Kadsura Japonica is the type of a genus consisting of seven or eight species, all natives of southern and western Asia, and its most northern member growing spontaneously in the southern islands and at the sea-level in Hondo as far north as the thirty-fifth degree of latitude. The flowers are not showy, but it is a plant of extraordinary beauty in the autumn, when the clusters of scarlet fruit are ripe, their brilliancy being heightened by contrast with the dark green lustrous persistent leaves. There is a fine specimen of this plant

MICHELIA COMPRESSA, Maxm.

in the garden attached to the Agricultural College at Tōkyō; but I have never seen it in any other, although it might well be grown wherever the climate is sufficiently mild to enable it to produce fruit.

Schizandra is familiar to American botanists as one species, the type of the genus, Schizandra coccinea, inhabits our southern states; in Japan two species occur, and one of these, Schizandra Chinensis, which grows also in Manchuria, carries the Magnolia family farther north than any of its other members. It is a vigorous plant, with long twining stems, and small unisexual white flowers, followed by clusters of brilliant red berry-like fruits, which in September and October enliven the forests of Hokkaido, where this plant is extremely common. Schizandra Chinensis is now well established in our gardens, flowering freely every year, although in the neighborhood of Boston it has not ripened its fruit.

The second species, Schizandra nigra, is much less common in Japan than Schizandra Chinensis, from which it may be distinguished by its broader leaves, larger flowers, and by its blue-black fruit and pitted seeds. It grows in southern Yezo, where, however, I failed to find it. Mr. Veitch collected it in September at Fukura, on the west coast of Hondo, and at the end of October I found a single plant near Fukushima, on the Nakasendō, in central Japan, from which I had the good fortune to gather a few ripe seeds; this interesting plant has not yet been brought under cultivation.

The forests of Japan are distinguished by the presence of a small family related to the Magnolias, and composed of three genera, Cercidiphyllum and Trochodendron, which are monotypic and endemic to Japan, and Euptelea, which has also a representative in the Himalaya forests. The plants of this genus are principally distinguished from the Magnolias by their generally diœcious minute flowers, which are either entirely destitute of a perianth or are furnished with a membranaceous four-lobed calyx. For this family the name of *Trochodendraceæ* has been proposed.[1]

Cercidiphyllum Japonicum is the most important tree of this family, and the largest and one of the most interesting deciduous trees of Japan, which more than any other of its inhabitants gives to the forests of Yezo their peculiar appearance and character (see Plate vi.). Here it inhabits the slopes of low hills and selects a moist situation in deep rich soil, from which the denseness of the forest and the impenetrable growth of dwarf Bamboos, which covers the forest-floor, effectually check evaporation. In such situations the Cercidiphyllum attains its greatest size, often rising to the height of a hundred feet, and developing clusters of stems eight or ten feet through. Sometimes it forms a single trunk three or four feet in diameter and free of branches for fifty feet above the ground; but more commonly it sends up a number of stems which are united together for several feet into a stout trunk, and then gradually diverge. The trunk of a typical Cercidiphyllum of this form appears in the frontispiece of this work; it is the reproduction of a photograph made on a hill near Sapporo, and represents a large but by no means an exceptionally large trunk, which at three feet above the ground girted twenty-one feet and six inches.

In Cercidiphyllum the leaves on sterile shoots are either alternate or opposite; in their axils small acute red buds, covered with four to six thin scarious slightly imbricated scales,

[1] Engler & Prantl, *Pflanzenfam.* iii. pt. ii. 21.

are formed early in the autumn. The branchlet ends during the winter in a small orbicular scar between two buds when the leaves are opposite, and at the side of a single bud when the leaves are alternate. Early in the following spring the buds develop short spur-like almost obsolete branches, which produce a single leaf and terminal flowers. Later a bud is formed in the axil of the leaf which, on fruit-bearing trees, appears between the leaf and the stalk of the fruit-cluster. The branches, therefore, in their second and third years, appear to be clothed with opposite or alternate leaves, although the leaves are in reality produced on lateral branches. The leaves are involute and coated on the lower surface in the bud with pale caducous pubescence, and are furnished with lanceolate acute caducous stipules slightly connate toward the base. The staminate and pistillate flowers are produced on separate individuals; the staminate are subsessile, solitary or fascicled, the pistillate solitary and pedunculate. The staminate flower is composed of a minute scarious calyx, divided to the base into four acute apiculate divisions, and of an indefinite number of stamens; the filaments are slender, elongated, and inserted on a conical receptacle; the anthers are oblong-lanceolate, attached at the base, apiculate by the prolongation of the narrow connective, and two-celled, the cells opening longitudinally throughout their length. The pistillate flower is composed of a membranaceous calyx divided into four unequal sepals laciniately cut on the margins, and of four or sometimes of five or six carpels inserted by their oblique bases on a prominent pyramidal receptacle; they are gibbous and acute on the ventral suture, and straight and rounded on the dorsal suture, and are gradually narrowed into elongated slender styles stigmatic on their inner face below the middle; the ovules are inserted in two rows on the placenta, and are descending and anatropous. The fruit is a cluster of two to six more or less spreading oblong stipitate follicles tipped with the persistent styles, and splitting through the ventral suture, which by a twist usually becomes external. The pericarp is thick, light brown, and lustrous, and separates into two layers; the outer layer is thin and membranaceous, and the inner layer is hard and woody, and lustrous on the inner surface. The seeds, which are closely imbricated in two rows, are pendulous, compressed, nearly square, attached obliquely, and covered with a thin, light brown membranaceous coat, which is produced into an elongated terminal wing three times as long as the body of the seed and slightly narrowed at the apex. The embryo is axile in copious fleshy albumen, with plane cotyledons about as long as the slender superior radicle turned toward the hilum.

The trunk of Cercidiphyllum Japonicum is covered with thick pale bark, deeply furrowed, and broken into narrow ridges. Similar bark covers the principal branches; these are very stout, and issuing from the stem nearly at right angles, gradually droop, the slender reddish branchlets in which they end being often decidedly pendulous. The upper branches and branchlets are erect, the whole skeleton of the tree showing, even in summer, through the sparse small nearly circular leaves, which are placed remotely on the branches; in the autumn the leaves turn clear bright yellow. In port and in the general appearance of its foliage, Cercidiphyllum, as it appears in the forests of Yezo, might, at first sight, be mistaken for a venerable Ginkgo-tree, which in old age has the same habit, with pendulous branches below and erect branches above; but the trunk and its covering are very different in the two trees.

Cercidiphyllum Japonicum is distributed from central Yezo southward through nearly the

CERCIDIPHYLLUM JAPONICUM, Sieb. et Zucc.

entire length of the Japanese islands. At the north it grows at the sea-level, and is very common, but on the main island it is confined to high elevations, and is rare. Except in Yezo, it seldom grows more than twenty or thirty feet high, and I never saw it, in Hondo, below 5,000 feet elevation, where, as at Yumoto, in the Nikkō Mountains, it is scattered through the lower borders of the Hemlock forest.

Cercidiphyllum Japonicum is a valuable timber-tree, producing soft straight-grained light yellow wood, which resembles the wood of Liriodendron, although rather lighter and softer, and probably inferior in quality. It is easily worked, and in Yezo is a favorite material for the interior finish of cheap houses and for cases, packing-boxes, etc. From its great trunks the Ainos hollowed their canoes, and it is from this wood that they make the mortars found in every Aino house and used in pounding grain. In New England, where there are now trees twenty feet high, Cercidiphyllum is very hardy, and grows rapidly; in its young state it is nearly as fastigiate in habit as a Lombardy Poplar, the trunk being covered from the ground with slender upright branches that shade it from the sun, which seems injurious to this tree, at least while young. As an ornamental plant, Cercidiphyllum is only valuable for its peculiar Cercis-like leaves, which, when they unfold in early spring, are bright red, and for its peculiar habit, as the flowers and fruit are neither conspicuous nor beautiful (see Plate vii.).

Of the other Japanese trees of this family, Euptelea polyandra is the least desirable as an ornamental plant, and it will probably never be very much cultivated except as a botanical curiosity. It is a small tree twenty to thirty feet in height, with a slender straight trunk covered with smooth pale bark, stout rigid chestnut-brown branchlets, marked with white spots, and wide-spreading branches, which form an open, rather unsightly head. The leaves are thin, prominently veined, bright green, sometimes five or six inches long and broad, nearly circular in outline, and deeply and very irregularly cut on the margins, with long, broad, apical points; they are borne on long slender petioles, and turn to a dull yellow-brown color before falling. The minute flowers appear in early spring before the leaves, and are produced in three or four-flowered clusters from buds formed early in the previous autumn. They have neither sepals nor petals, and consist of a number of slender stamens surrounding the free clustered carpels. The fruit, which ripens in November, is not more showy than the flowers; it is a small stalked samara half an inch long, and furnished with an oblique marginal membranaceous wing. The handsomest thing about this tree is the winter-bud, which is obtuse, half an inch long, and covered with imbricated scales, which are bright chestnut-brown, and as lustrous as if they had been covered with a coat of varnish. Euptelea polyandra is found in the mountainous forests of central Japan, usually on the banks or in the neighborhood of streams between 2,000 and 3,000 feet above the sea-level, but does not appear to be anywhere very common.

The third genus of this family, Trochodendron, like Euptelea, produces flowers without sepals and petals. The only species, Trochodendron aralioides, is a small handsome glabrous evergreen tree, with alternate broadly rhomboidal crenulate penniveined leaves, four or five inches long; they are borne on elongated stout petioles, and are clustered at the extremities of the branches. The flowers are produced in short terminal racemes, and consist of numerous anthers raised on slender filaments and surrounding the carpels, which are connate in a vertical series, and ripen into a small fleshy drupe crowned with the remnants of the persistent styles. Trochodendron is a tree from fifteen to twenty-five feet in height, and is said to be very common in some parts of the country, although I never saw it growing wild; and it is certainly not an inhabitant of the alpine forests or of Hokkaido, as stated in some works on the Japanese flora, although, perhaps, it occurs in northern Hondo at the sea-level, as it is hardy in the gardens of Nikkō at an elevation of 2,000 feet above the ocean. Trochodendron aralioides is often cultivated by the Japanese, and fine specimens of this tree are found scattered through public and private gardens in Tōkyō and Yokohama.

PLATE VII.

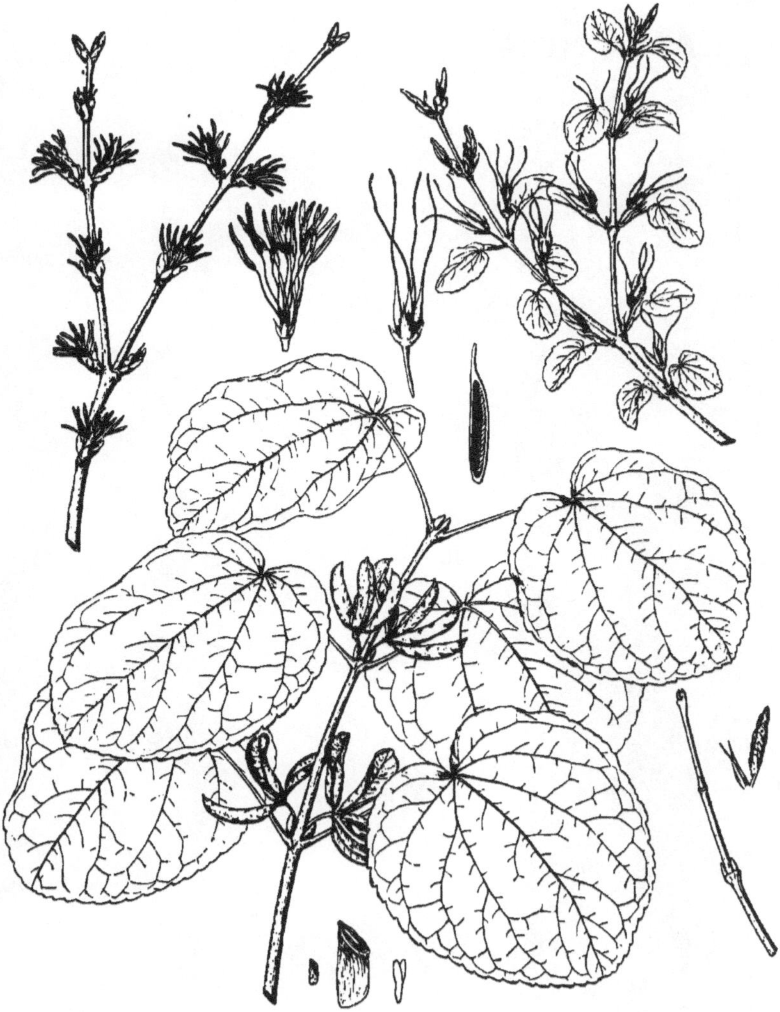

C. E. Faxon del.

CERCIDIPHYLLUM JAPONICUM, Sieb. et. Zucc.

THE TERNSTRŒMIA, LINDEN, AND RUE FAMILIES.

The chiefly tropical family, Ternstrœmiaceæ, which, in North America, is represented only by Gordonia and Stuartia, trees and shrubs of the southern states, in Japan appears in eight genera, in which are a number of interesting plants, although none of them become very large trees. Of these, Camellia Japonica is horticulturally the most important, for its relative, Camellia theifera, the Tea-plant, is evidently a Chinese or Assam introduction, and not a native of Japan. In southern Japan the Camellia is a common forest-plant from the sea-level to an altitude of 2,500 feet, on the east coast growing as far north as latitude thirty-six, and nearly two degrees farther on the west coast. Here it is a dwarf bush only two or three feet high, although where the soil and climate favor it, the Camellia becomes a tree thirty or forty feet tall, with a handsome straight trunk a foot in diameter, covered with smooth pale bark hardly distinguishable from that of the Beech. In its wild state the flower of the Camellia is red, and does not fully expand, the corolla retaining the shape of a cup until it falls. In Japan, certainly less attention has been paid to the improvement of the Camellia than in Europe and America, although double-flowered varieties are known; and as an ornamental plant it does not appear to be particularly popular with the Japanese; it is sometimes planted, however, in temple and city gardens, especially in Tōkyō, where it is not an uncommon plant, and where beautiful old specimens are to be seen.

Tsubaki, by which name Camellia Japonica is known in Japan, is more valued for the oil which is pressed from its seeds than for the beauty of its flowers. This oil, which the other species of Camellia also produce, is used by the women in dressing their hair, and is an article of much commercial importance. The wood of Camellia is close-grained, moderately hard, and light-colored, turning pink with exposure; it is cut into combs, although less valued for this purpose than boxwood, and is manufactured into numerous small articles of domestic use. Sasan-kuwa, Camellia Sasanqua, a small bushy tree of southern Japan and China, is perhaps more commonly encountered in Japanese gardens than the Tsubaki, and in the first week of November it was just beginning to open its delicate pink flowers in the gardens of Nikkō, although the night temperature was nearly down to the freezing point.

Ternstrœmia Japonica and Cleyera ochnacea are small bushy trees scattered from India to southern Japan, where they are considered sacred by votaries of the Shintō religion, and are therefore planted in the grounds of Shintō temples and in most private gardens. The evergreen foliage of these two plants is handsome, especially that of Ternstrœmia, but the flowers and fruit possess little beauty, and they owe their chief interest to their association with Japanese civilization.

Eurya Japonica is another member of the family, of wide range from Ceylon and India to China, the Fejee Islands, and Japan, where it is exceedingly common in the southern islands and in Hondo as far north, at least, as the Hakone Mountains. It is usually a shrub only a few feet high; but I saw a specimen in the woods surrounding a temple near Nakatsu-gawa, on the Nakasendō, which was fully thirty feet in height, with a well-formed trunk nearly a foot in diameter. Eurya, although not particularly handsome, is interesting from the color of the leaves, which are yellowish green on the upper surface and decidedly yellow below.

Stuartia is represented in eastern America by two handsome shrubs, one an inhabitant of the coast region of the south Atlantic states, and the other of the southern Alleghany Mountains; in Japan there are two and, perhaps, three species. Of these, Stuartia monadelpha, which inhabits also central China, appears to be a southern plant only; at any rate, I saw nothing of it in Japan, nor of the little known Stuartia serrata of Maximowicz. The third species, Stuartia Pseudo-Camellia, is common on the Hakone and Nikkō Mountains between 2,000 and 3,000 feet elevation, where it is a most striking object from the peculiar appearance of the bark; this is light red, very smooth, and peels off in small flakes like that of the Crape Myrtle (Lagerstrœmia); to this peculiarity it owes its common name, Saru-suberi, or Monkey-slider. Stuartia Pseudo-Camellia is often a tree of considerable size; on the shores of Lake Chuzenji we measured a specimen whose trunk at three feet from the ground girted six feet, and which was upward of fifty feet high; and specimens nearly as large are common on the road between Nikkō and Chuzenji. The flowers of this tree, which resemble a single white Camellia, are smaller and less beautiful than the flowers of our coast species, Stuartia Virginica, but they are larger than those of the second American species, Stuartia pentagyna, a handsome plant, which is not made enough of in our northern gardens, where it is perfectly hardy and one of the best of the summer-flowering shrubs. Stuartia Pseudo-Camellia was sent to America nearly thirty years ago by the late Mr. Thomas Hogg, and it appears to have flowered in the neighborhood of New York several years before it was known in Europe, where of late it has attracted considerable attention.[1] In New England this Japanese species appears perfectly hardy, and two years ago it flowered in the Arnold Arboretum.

Stachyurus præcox, another Japanese member of this family, is still little known in our gardens, although it was one of the plants sent by Mr. Hogg to New York soon after the opening of Japan to foreign commerce. It appears hardy in the neighborhood of New York, as there is at least one plant established in Prospect Park, on Long Island. In Japan Stachyurus is exceedingly common in the mountain forests and at the sea-level from southern Yezo to Kyūshū, appearing as a tall graceful shrub, with thin semiscandent branches and ovate-lanceolate acute leaves. In summer or early autumn it forms axillary spikes of flower-buds two or three inches long, and in very early spring, before the appearance of the leaves, these buds expand into bell-shaped pale yellow flowers; these are not more than a third of an inch long, but they are produced in great profusion, and as they appear so early in the season Stachyurus will probably prove a popular plant if it is found to flourish in cultivation. The genus is represented in central China and in the Himalaya Mountains with a second species described as a small tree.

The genus Actinidia, woody climbers of the Himalayas and eastern Asia, appears in Japan in three species, of which two, at least, are exceedingly common and conspicuous features of the mountain vegetation. Of these, the largest and most common, especially at the north,

[1] See *Rev. Hort.* 1879, 430, t. — *Gard. Chron.* ser. 4, iv. 188, f. 22. — *Bot. Mag.* cxv. t. 7045.

is Actinidia arguta; little need be said of this handsome plant, as it is now common and well established in our gardens, where it grows with great vigor and rapidity, and where it is one of the best plants of its class. We have heard a good deal of the value of the fruit of this plant, which is depressed-globular, an inch across, and greenish yellow; it is eaten in Japan, but the flavor is insipid, and its merits appear to have been exaggerated. It was offered for sale in the streets of Hakodate in great quantities, but, of course, green and hard, as the Japanese use all their fruit before it ripens.

Actinidia polygama, although it inhabits Manchuria and Saghalin, and is common in the forests of Hokkaido, is more abundant in those which cover the mountains of central Japan; it is a slenderer plant than Actinidia arguta, with elliptical acute slightly serrate long-stalked leaves. The fruit is an inch and a half long, half an inch broad in the middle, and narrowed at both ends; it is canary-yellow, rather translucent, soft and juicy, with an extremely disagreeable flavor. Actinidia polygama does not, like Actinidia arguta, climb into the tops of tall trees; its weaker stems tumble about and form great tangles, sometimes twenty feet or more across, and fifteen or twenty feet high. The most remarkable thing about this plant is that in summer the leaves toward the ends of the branches become pale yellow, either over their entire surface or only above the middle, not because they are drying up or ripening, but apparently from an insufficient supply of chlorophyll. The effect that the plants produce at this time is curious and interesting, and when seen from a distance growing on a mountain-side or on the banks of a stream, they appear like huge bushes covered with pale yellow flowers. This fine plant is still little known in cultivation, but if it flourishes in New England like Actinidia arguta it will form a most valuable addition to our shrubberies.

Actinidia Kolomikta, which is found also in Manchuria and northern China, is much less common in Japan than the other species. I saw it only on the rocky cliffs of a hill near Sapporo, where it was growing with Rhododendrons and Menziesia, and where it was a delicate, slender vine, with stems only a few feet in length. Unfortunately, there were no seeds to be obtained, and I am doubtful if this species has ever been introduced into our gardens, although the name often appears in nurserymen's catalogues.

In the forests of Japan are found two Lindens. They are both extremely common in Yezo, but in the other islands are rare, and confined to mountain-slopes of considerable elevation. The larger of the two, Tilia Miqueliana (see Plate viii.), is a handsome tree, often growing in central Yezo to the height of one hundred feet, and forming a trunk four or five feet in diameter. As it is only seen crowded among other trees in the forest, the branches are short, and the head is oblong and rather narrow. The bark, like that of all the Lindens, is broken by longitudinal furrows, and is light brown or dark gray. The young branchlets are unusually stout for a Linden-tree, and in their first season are covered, as are the large ovate-obtuse winter-buds, with hoary tomentum. The leaves are deltoid or deltoid-obovate, abruptly contracted at the apex into broad points, obliquely truncate or subcordate at the base, and coarsely and sharply serrate with incurved callous teeth; they are four to six inches long, three or four inches broad, rather light green, and more or less puberulous on the upper surface, and pale and tomentose on the lower, especially on the prominent midribs and primary veins, in their axils, and on the stout petioles, which are two or three inches in length. The peduncle-bract is rounded at the apex, sessile or short-stalked, three or four inches long, from one third to two thirds of an inch broad, and, like the slender stems and branches of the flower-cluster and its bractlets, covered with pale tomentum. The flowers appear in Sapporo toward the middle of July. Like those of the American Lindens and of two species of eastern Europe, Tilia petiolaris and Tilia argentea, which Tilia Miqueliana resembles in several particulars, the flowers are furnished with petal-like scales, to which the stamens, united in clusters, are attached. The sepals are ovate-acute, tomentose on the two surfaces, especially on the inner, and shorter than the narrow obovate petals. The style, like the stamens, is longer than the petals, and is coated at the base with thick pale hairs, which also cover the ovary. The fruit, which ripens in October, is ovate to oblong, wingless, and nearly half an inch long. It is from the inner bark of this species that the Ainos make their ropes.

Tilia Miqueliana [1] is comparatively little known, having at one time been confounded with Tilia Mandshurica, which does not reach Japan. This noble tree will probably thrive in the northern states, as plants which have been growing for a few years in the Arnold Arboretum appear perfectly hardy. In Europe it is cultivated as Tilia Mandshurica (Kew), and as Tilia heterophylla (Paris), although it does not appear to be much better known there than it is in the United States.

The second Japanese Linden is a small tree, rarely growing more than fifty or sixty feet tall in Hokkaido, where, perhaps, it is rather less abundant than Tilia Miqueliana. In books it appears as Tilia cordata, var. Japonica; but Tilia cordata is a synonym for Tilia ulmifolia, a common European and north Asian species, so that unless the Japanese plant is found specifically distinct, which is not probable, it should be known as Tilia ulmifolia, var. Japonica. It is a round-headed tree with dark brown bark, slender red-brown branches, glabrous, like the buds, even when young, and marked with oblong pale lenticels. The leaves are broadly ovate or nearly orbicular, contracted at the apex into short or long broad points, and usually cordate, or occasionally oblique, at the base, and sharply serrate with incurved callous teeth; they are membranaceous, light green and lustrous on the upper surface, light green, pale, or nearly white on the lower surface, which is marked by conspicuous tufts of rufous hairs in the axils of the principal veins, three or four inches long, and two or three inches broad. The peduncle-bract is from three to three and a half inches long, and half an inch broad, with a slender stalk sometimes an inch in length. The stem and branches of the flower-cluster are slender and glabrous. The sepals, which are acute, slightly puberulous on the outer surface, ciliate on the margins, and furnished on the inner surface at the base with large tufts of pale hairs, are shorter than the narrow acute petals; the ovary is clothed with white tomentum. The fruit is oblong, or slightly obovate, and covered with rusty tomentum. The petaloid-scales, which Maximowicz [2] found developed in some of the flowers of this tree, I have not seen.

This is the only Linden cultivated by the Japanese, who occasionally plant it in temple gardens, especially in the interior and mountainous part of the empire. It was introduced in

---

[1] Tilia Miqueliana, Maximowicz, *Mél. Biol.* x. 585.
Tilia Mandshurica, Miquel, *Prol. Fl. Jap.* 206 (in part).—Franchet & Savatier, *Enum. Pl. Jap.* i. 67 (in part).
[2] *Mél. Biol.* x. 585.

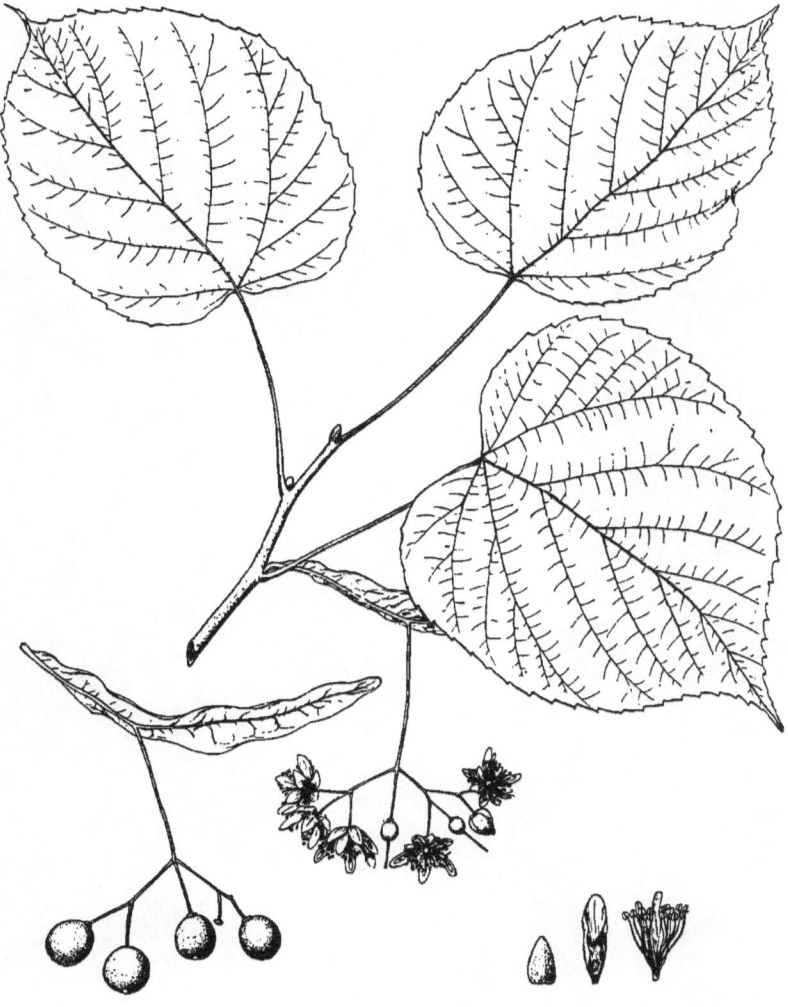

TILIA MIQUELIANA, Maxm.

1886 into the Arnold Arboretum, and has so far proved hardy. It is, however, scarcely distinct enough from the European plant to make its cultivation as an ornamental tree particularly desirable.

Elæocarpus, a genus of the Linden family distributed with many species through tropical Asia, Australia, and the Pacific islands, is represented in Japan by two fine trees, found only in the extreme southern part of the empire. Of these I saw only Elæocarpus photiniæfolia, a noble tree, planted in the gardens of a temple in the seashore town of Atami, where there are the largest Camphor-trees in Japan and good specimens of a number of other southern trees.

The Rue family has a number of woody plants in Japan. Of these, Skimmia Japonica is the only one which is much known in our gardens, although Phellodendron Amurense and Orixa Japonica are now found in most large botanical collections. Evodia rutæcarpa, a shrub or very small tree, with large, pinnate, strong-smelling leaves and terminal heads of minute flowers, although not at all handsome, is an interesting plant, as it is from the bark that the Japanese obtain the yellow pigment which they use in dyeing. For this purpose the bark, which, with the exception of its thin brown outer coat marked with pale lenticels, is the color of gamboge, is torn off in long strips, air-dried, and sent to the large cities. Evodia rutæcarpa, which also inhabits central China and the Himalayas, is now becoming rare in Japan, and I saw it only on the coast near Atami; it is said to be still abundant, however, in Aidzu and on the peninsula of Yamato. The scarlet aromatic fruit is used by the Japanese in medicine.

In Xanthoxylum there are four Japanese species. Of these, the most common, and the most widely distributed at the north and in the mountainous regions of the main island at elevations of from 2,000 to 4,000 feet above the sea, is Xanthoxylum piperitum. It is a bushy shrub with many slender stems, or rarely a small tree with a well-developed trunk three or four inches in diameter; it is always a handsome plant, with dark or often nearly black branchlets marked by pale spots and armed with stout straight spines; with narrow unequally pinnate leaves of about six pairs of ovate pointed leaflets, very dark green on the upper surface and pale on the lower; with small inconspicuous flowers and with heads of handsome showy fruit four to six inches across; the pods are rusty brown; and the seeds, which do not drop for some time after the pods open, are black and lustrous. The fruits of this plant are gathered in large quantities by the Japanese before the pods open, and are used as a condiment and in cooking, as we use pepper. In Hakodate and other northern towns it is commonly exposed for sale throughout the year.

A nobler plant than Xanthoxylum piperitum, and certainly one of the most beautiful of the genus, is Xanthoxylum ailanthoides, which I saw only on the Hakone Mountains, where it is abundant, and near the coast at Atami. It is a round-topped broad-branched tree, sometimes fifty or sixty feet tall, with a trunk twelve to eighteen inches in diameter, covered with pale bark, upon which the corky excrescences common in many species of this genus are well developed. The branchlets are stout and pale, and are covered with short stout spines. The leaves vary from eighteen inches to four feet in length, and are unequally pinnate, with about ten pairs of lateral leaflets and stout red-brown petioles; the leaflets are dark green, and conspicuously marked on the upper surface with oil-glands, pale or nearly white on the lower surface, ovate-acute, often slightly falcate, long-pointed, rounded or subcordate at the base, finely serrate, stalked, four to six inches long and two to four inches broad. The flowers, which are greenish white and small, and inconspicuous like those of all the plants of this genus, appear on the Hakone Mountains at the end of August or early in September in clusters four or six inches across. The fruit I have not seen. In habit and in foliage this is one of the most beautiful trees which I saw in Japan; but as it does not range far north or ascend to the high mountains, it is not probable that it will prove hardy in our northern states. The other Japanese species of Xanthoxylum are shrubs of no great beauty or interest.

Simarubæ, a mostly tropical family, to which the familiar Ailanthus of northern China belongs, appears in Japan only in Picrasma quassioides, a member of a small tropical Asian genus, which, as an inhabitant of Yezo, seems to have strayed far beyond the limits of its present home. As Picrasma quassioides appears in the forests near Sapporo, it is a slender tree twenty to thirty feet in height, with a trunk about a foot in diameter. The branchlets are stout, dark red-brown, and conspicuously marked by pale lenticels. The leaves are unequally pinnate, with slender reddish petioles and four or five pairs of lateral leaflets, which increase in size from the lower pair to the uppermost; they are membranaceous, very bright green, ovate-acute, finely serrate, stalked, three to five inches long, and an inch to an inch and a half broad. The flowers, which are produced in loose, long-branched, few-flowered, axillary clusters, are yellow-green, and not at all showy; but the drupe-like fruit is bright red and handsome in September, when the thickened branches of the corymb are of the same color. It is, however, for the beauty of the color of its autumn foliage that Picrasma quassioides should be brought into our gardens. The leaves turn early, first orange and then gradually deep scarlet, and few Japanese plants which I saw are so beautiful in the autumn as this small tree, which, judging from its northern home in Japan, may be expected to flourish in our climate. It is a plant of wide distribution, not only in Yezo and Hondo, but in Corea and in northern and central China; it occurs on Hongkong and Java, and is common on the subtropical Himalayas, which in Garwhal it ascends to an elevation of 8,000 feet above the ocean. To the bitterness of the inner bark, which in this particular resembles that of the Quassia-tree of the same family, it owes its specific name.

THE HOLLY, EVONYMUS, AND BUCKTHORN FAMILIES.

JAPAN and eastern North America are equally rich in species of Holly, there being thirteen or fourteen in each of the two regions. In Japan, however, Hollies grow to a larger size than they do in North America, there being eight or nine trees in this genus in the Mikado's empire, and only four in the United States; and some of the Japanese Hollies are much larger and far more beautiful than any of our species. The most beautiful of them all is certainly the southern Ilex latifolia, an evergreen tree now occasionally seen in the gardens of southern Europe, where it was first carried more than fifty years ago. Although a native of southern Japan, Ilex latifolia appears perfectly at home in Tōkyō, where it is often seen in large gardens and temple grounds, and where it occasionally makes a tree fifty to sixty feet in height, with a straight tall trunk covered with the pale smooth bark which is found on the stems of most plants of this genus. The leaves are sometimes six inches long and three or four inches broad, and are very thick, dark green, and exceedingly lustrous. The large scarlet fruit of this tree, which does not ripen until the late autumn or early winter months, and which is produced in the greatest profusion in nearly sessile axillary clusters, remains on the branches until the beginning of the following summer. Ilex latifolia is probably the handsomest broad-leaved evergreen tree that grows in the forests of Japan, not only on account of its brilliant abundant fruit, but also on account of the size and character of its foliage. It may be expected to prove hardy in Washington, and will certainly flourish in the southern Atlantic and Gulf states.

Ilex integra is also a beautiful and distinctly desirable ornamental tree, often cultivated in the temple gardens of Japan, where it frequently reaches a height of thirty or forty feet. The leaves are narrow, obovate, three or four inches long, and apparently quite entire. The fruit, which is rather long-stalked, is nearly half an inch in diameter, and very showy during the winter. A variety of this species (var. leucoclada, Maximowicz), a shrub two to three feet high, with narrower leaves and smaller fruit, is a northern form, growing as far north as southern Yezo. On Mount Hakkoda, near Aomori, we found this plant in full flower and with ripe fruit on the 2d of October, and secured a supply of the seeds, so that its hardiness can be tested in the northern states. It must be remembered, however, that, although this plant, and several other broad-leaved evergreen shrubs, including two or three species of Holly, grow in Japan in a higher latitude than Massachusetts, they are protected, as Maximowicz has already pointed out, during the winter by an undisturbed covering of snow, and are not exposed, therefore, to the changes of climate which endanger the existence of many plants in eastern America. In Japan, moreover, plants do not suffer from the summer and autumn droughts, which often sap their vitality in the United States, and are often more directly responsible for the apparent want of hardiness of many plants than intense winter cold.

A third Japanese evergreen species, Ilex rotunda, is also occasionally cultivated by the

Japanese, although I saw only two or three specimens of this plant; these were handsome trees, thirty to forty feet in height, with well-formed trunks twelve to thirteen inches in diameter. The leaves are broadly ovate to nearly orbicular, with entire thickened margins, and are very dark green and lustrous, although not thick nor very coriaceous. The fruit is smaller than that of the two species already mentioned, and rather oblong in outline.

A very distinct evergreen species, Ilex pedunculosa, is exceedingly common on the Nakasendō, the great central mountain road of Japan, in the valley of the Kisogawa. This plant is sometimes a shrub two or three feet in height, and is sometimes twenty or thirty feet high, when it is a well-formed tree, with a narrow round-topped head. The leaves are lustrous, two to three inches long, ovate-acute, entire, and long-petiolate. The stems of the flower-clusters, from which is derived its specific name and which are longer than the leaves, hold the large bright-red fruit, which is solitary, or arranged in clusters of three or four, well outside the leaves, giving to the plants a peculiar and beautiful appearance in the autumn. Occasionally a tree of this species was seen in the garden of an inn on the Nakasendō; but it is evidently little known or cultivated in Japan, and apparently has not been introduced into western gardens. Ilex pedunculosa will certainly flourish in western and southern Europe, and I am not without hope that it will survive and possibly thrive in the northern United States, as in Japan it is found at high elevations in a region of excessive winter cold.

Ilex crenata is the most widely distributed and the most common of the Japanese Hollies with persistent leaves; this plant is abundant in Hokkaido, on the foothills of Mount Hakkoda, and on the sandy barrens near Giffu, on the Tōkaidō; and I encountered it in nearly every part of the empire which I visited. It is usually a low much-branched rigid shrub, three or four feet high; but in cultivation it not infrequently rises to the height of twenty feet, and, assuming the habit of a tree, is not unlike the Box in general appearance. The leaves, which are light green and very lustrous, vary considerably in size and shape, although they are rarely more than an inch long, and are usually ovate-acute, with slightly crenate-toothed margins. The black fruit is produced in great profusion, and in the autumn adds materially to the beauty of the plant. This is the most popular of all the Hollies with the Japanese; and a plant usually cut into a fantastic shape is found in nearly every garden. Varieties with variegated leaves are common and apparently much esteemed. Ilex crenata and several of its varieties with variegated foliage were introduced into western gardens many years ago and are occasionally cultivated, although the value of this plant as an under-shrub appears to be hardly known or appreciated outside of Japan. Of the broad-leaved Japanese evergreens, I have the most hope of success with Ilex crenata in this climate; and if it proves really hardy it will be a most useful addition to our shrubberies.

Ilex Sugeroki, another evergreen species quite unknown, I believe, in gardens, may be expected to thrive in Europe, and possibly in the northern United States, as it is an inhabitant of southern Yezo and northern Hondo, where on Mount Hakkoda we found it in fruit, and were able to secure a supply of the seeds. It is a spreading bush five or six feet high, with stout branchlets, light green ovate leaves an inch long, rounded at the apex and coarsely crenulate-toothed above the middle, and with bright scarlet long-stalked solitary fruit half an inch in diameter. Ilex Sugeroki is an unusually handsome plant in the autumn, and of considerable horticultural promise.

Of the section of the genus with deciduous leaves (Prinos), represented in eastern North America by the familiar Black Alder (Ilex verticillata) of our northern swamps and by the arborescent Ilex Monticola of the Alleghany Mountains, there are several species in Japan. The largest of these, Ilex macropoda, is a widely distributed, but not a common plant. I saw it on the cliffs at Mororan on the shores of Volcano Bay, on the hills above Nikkō, and on the flanks of Mount Koma-ga-take in central Japan, although only a single plant in each of these widely separated localities. Ilex macropoda is a round-headed tree, twenty to thirty feet in height, with a trunk sometimes a foot in diameter. It is a well-shaped handsome tree, with stout branchlets furnished with short lateral spurs and ample, membranaceous, ovate-acute, long-petioled leaves conspicuously reticulate-veined, which turn bright clear yellow in the autumn, when they make a beautiful contrast with the bright red long-stalked fruit, which, although not very large, is exceedingly abundant. Ilex macropoda grows not only far north, as Professor Miyabe has recently written me of its discovery in the neighborhood of Sapporo, but in the most exposed situations and at high elevations; and there is no reason, therefore, why it should not thrive in our northern states, where it may be expected to add considerably to the beauty of shrubberies in the autumn and early winter.

A much more common plant than Ilex macropoda is Ilex Sieboldii, although this species does not reach Hokkaido or ascend to high elevations on the mountains of Hondo. It much resembles our North American Ilex verticillata and Ilex lævigata, although much less beautiful than either of these species, the fruit being smaller and less highly colored. Ilex Sieboldii is a tall spreading shrub, very common in low grounds and near the borders of streams, with slender stems often twelve or fifteen feet tall, small ovate-acute sharply serrate conspicuously veined leaves, and small scarlet fruit clustered on the short lateral spur-like branchlets. In the autumn the leafless branches of this shrub covered with fruit are sold in immense quantities in the streets of Tōkyō for the decoration of dwelling-houses, for which purpose they are admirably suited, as the berries remain on the branches and retain their color for a long time. Ilex Sieboldii was introduced many years ago into American gardens by the late Thomas Hogg; it is an old inhabitant of the Arnold Arboretum, where it now flowers and produces its fruit every year. As an ornamental plant, however, it is less desirable than the related American species, and it will probably only be cultivated in this country and in Europe as a curiosity, or in botanic gardens.

The other Japanese Hollies with deciduous leaves, Ilex serrata, which is closely related to and resembles Ilex Sieboldii, and Ilex geniculata, a rare shrub of the high mountains, with black fruit, I was not fortunate enough to find.

The flora of Japan is rich in Evonymus, there being no less than nine species found within the limits of the empire. Of these the best known in our gardens is the evergreen Evonymus Japonicus, now cultivated in all temperate countries, and its climbing variety usually known as Evonymus radicans. Evonymus Japonicus is a small tree generally distributed at low elevations, and more common in the south than at the north, although it grows naturally in the cold climate of southern Yezo, where, however, it does not attain a large size, and where its presence may be accounted for by the thick covering of snow which protects it in winter. The scandent variety is a hardier plant found carpeting the ground under the forests of

Hokkaido, and in the mountain regions of Hondo climbing high on the trunks of trees, which it encircles with great masses of lustrous foliage borne on stout branches standing out at right angles sometimes to the length of several feet; the leaves vary from an inch to four or five inches in length and correspondingly in width, and show the connection of the climbing plant with the arborescent type.

There is a second arborescent Evonymus in Japan, a variety of the widely distributed and variable Evonymus Europæus, to which the name var. Hamiltonianus is given. This handsome plant, with stout branchlets, large leaves, and showy fruit, was introduced from Japan several years ago by the late Thomas Hogg, and it is now well established in the Arnold Arboretum, where it flowers and fruits freely. It is one of the commonest of the Japanese species in all mountain regions, and grows at least as far north as central Yezo, where it becomes a tree twenty to thirty feet in height.

Evonymus alatus, a variable plant in the development of the wings on the branches, to which it owes its specific name, and in the size of the leaves and fruit, in some of its forms, is also very abundant in the north and on the mountains of central Japan. The wing-branched variety, which is the only deciduous-leaved Evonymus which I saw in Japanese gardens, where it is rather a favorite, is now well known in those of the United States and of Europe, where it is valued for the peculiar pink color the leaves assume in very late autumn. The variety subtriflora, a more northern plant, with slender terete branchlets and small fruit, is, I believe, unknown in gardens. It is one of the commonest shrubs in the mountain forests of Japan, and on the shores of Lake Chuzenji, in the Nikkō Mountains, I saw it rising to the height of fifteen or eighteen feet, with slender diverging stems.

In northern Japan there are three other species of Evonymus, all tall shrubs, with large leaves and large showy fruit suspended on long slender stalks, which may be expected to thrive in our climate, and to be decided acquisitions in our shrubberies. Of these Evonymus Nipponicus and Evonymus oxyphyllus produce globose fruit, and Evonymus macropterus more or less broadly winged fruit.

Of Celastrus nothing need here be said of the now well-known Celastrus articulatus, which is one of the commonest plants on the mountains of Japan, except that its leafless branchlets, covered with fruit, are sold in the autumn in great quantities in all Japanese towns, where they are used in house decoration, for which purpose they are admirably suited, as the bright-colored fruit remains on them for many weeks. The second Japanese species, Celastrus flagellaris, I saw only in the Botanic Garden in Tōkyō, where there is a single small plant; it is a common Manchurian species, but appears to be exceedingly rare in Japan. I judge that it has no particular horticultural value.

Half a dozen genera of Rhamnaceæ are included in the flora of Japan, among them Zizyphus, perhaps an introduced plant, often cultivated as a fruit-tree; Berchemia racemosa, a twining shrub with long slender branches, very ornamental during the last weeks of summer, when the half-ripened fruit, which is produced in large terminal clusters, is bright red; two or three species of Rhamnus, of no horticultural value, and the curious tree, Hovenia dulcis, an inhabitant also of China and the Himalaya region, and in Japan often cultivated for the thickened sweetish fruit-stalks, which are edible, although insipid in flavor, and which enjoy among the Japanese a certain reputation for curative properties. Hovenia was first introduced into Europe eighty years ago, and is occasionally seen in the gardens of southern France and Italy and of our middle states. In general appearance this tree, which is sometimes thirty or forty feet in height, is not unlike a large-leaved Pear-tree, and as an ornamental plant possesses little value.

THE MAPLE FAMILY.

In arborescent plants of the family of Sapindaceæ, Japan is richer than eastern America, owing to the multiplication of species of Maple in the former country. Æsculus, on the contrary, which finds its headquarters in North America, where there are five species, appears in Japan in only one, — Æsculus turbinata. This, however, is a noble tree, — one of the largest and stateliest of all Horse-chestnuts. In the forests of the remote and interior mountain regions of central Hondo, at elevations between 2,000 and 3,000 feet, Horse-chestnuts, eighty to one hundred feet tall, with trunks three or four feet in diameter, are not uncommon. These were perhaps the largest deciduous trees which I saw on the main island growing naturally in the forest, that is, which had not been planted by men, and their escape from destruction was probably due to their inaccessible position and to the fact that the wood of the Horse-chestnut is not particularly valued by the Japanese. In habit and in the form, venation, and coloring of the leaves, the Japanese Horse-chestnut resembles the Horse-chestnut of our gardens, the Grecian Æsculus Hippocastanum, and at first sight it might easily be mistaken for that tree, but the thyrsus of flowers of the Japanese species, which is ten or twelve inches long and only two and a half to three inches broad, is more slender; the flowers are smaller and pale yellow, with short, nearly equal petals ciliate on the margins; and the fruit is that of the Pavias, being smooth and showing no trace of the prickles which distinguish the true Horse-chestnuts. The Japanese Horse-chestnut reaches southern Yezo, finding its most northern home near Mororan, on the shores of Volcano Bay, at the level of the ocean; it is generally distributed through the mountainous parts of the three southern islands, sometimes ascending in the south to an elevation of 4,000 or 5,000 feet. There seems to be no reason why this tree, which has already produced fruit in France, should not flourish in our northern states, where, as well as in Europe, it is still little known. In northern Japan the fruits are exposed for sale in the shops, although they are probably used only as playthings for the children.

To the Maples the forests of Japan owe much of their variety, beauty, and interest. Not less than twenty species are known in Japan, while in all of North America there are only nine, with six on the eastern side of the continent. None of the Japanese Maples, however, grow to the size of real timber-trees, or can be compared in massiveness and grandeur with some of the American species, which are unrivaled in size and beauty by the Maples of any other part of the world.

Some of the Japanese Maples are exceedingly common and form a conspicuous feature of the forest vegetation, and others are rare and confined to comparatively small regions. Several of the species I did not see at all, and of others only one or two isolated individuals. The most common of the Japanese Maples, and the largest, is Acer pictum, a handsome small tree, not unlike our Sugar Maple in general appearance; it is one of the most abundant trees in the forests of Hokkaido, where it occasionally attains the height of fifty feet, and forms a trunk eighteen inches in diameter. It is a tree of wide and general distribution in Japan, Manchuria, China, and northern India, and even in Japan varies remarkably in the size and pubescence of the five to seven-lobed leaves truncate at the base, and in the size and shape of the fruit. This tree must be extremely beautiful in May, when the yellow flowers are just opening, for the large lengthened inner scales of the winter-buds are then bright orange-color and very showy. The autumn coloring of the leaves I did not see; it is described as yellow and red.

Of more interest to the lovers of novelties is Acer Miyabei (see Plate ix.), the latest addition to the list of Japanese Maples. It is a tree thirty to forty feet in height, with a trunk twelve to eighteen inches in diameter, covered with pale deeply furrowed bark, spreading branches which form a round-topped handsome head, and stout branchlets orange-brown in their first, and ashy gray in their second season. The leaves are five-lobed by narrow sinuses, with acute entire irregularly two to three-lobed divisions, and are cordate or almost truncate at the base, five-ribbed, conspicuously reticulate-veined, puberulous on the ribs and in their axils on the upper surface, and more or less covered with ferrugineous pubescence on the lower surface, especially on the ribs and veins; they are dark green above, pale below, and four or five inches long and broad, and are borne on stout petioles enlarged at the base, two to seven inches in length, and thickly coated while young with pale hairs, which also cover the unfolding leaves. The flowers, which are yellow, are produced on slender pedicels in few-flowered, short-stalked corymbs. The sepals and petals are narrow, obovate, acute and ciliate on the margins; in the male flowers the stamens, composed of filiform filaments and minute ovate anthers, are inserted between the lobes of a conspicuous disk, and are longer than the petals; the pistil is minute and rudimentary; in the fertile flowers the stamens are rudimentary and shorter than the ovary, which is coated with long white hairs. The style, which is described as somewhat shorter than the revolute stigmas, is caducous. The fruit is two inches long, with broad puberulous nutlets diverging at right angles to the stem, and thin, slightly falcate, conspicuously veined wings. This fine tree, which is closely related to the European Acer platanoides, was discovered a few years ago in the province of Hidaka, in Hokkaido, by Professor Kingo Miyabe, the accomplished professor of botany in the college at Sapporo and the author of an important work on the flora of the Kurile Islands, in whose honor it was named in 1888 by Maximowicz.[1]

On the 18th of September we stopped quite by accident to change cars at the little town of Iwanigawa, a railroad junction in Yezo, some forty or fifty miles from Sapporo, and, having a few minutes on our hands, strolled out of the town to a small grove of trees in the hope that they might prove interesting. In this grove, occupying a piece of low ground on the borders of a small stream, and chiefly composed of Acer pictum, our Japanese guide recognized at a glance a number of fine trees of Acer Miyabei covered with fruit, and surrounding the house of an officer of the Imperial Forest Department, who had been living for years in entire ignorance of the fact that he was enjoying the shade of one of the rarest trees in Japan. The find was a lucky one, for Iwanigawa is a long way from the station where this species

[1] *Mél. Biol.* xii. 725.

had been discovered, and mature fruit had not been seen before; and from these trees I obtained later from Professor Miyabe a supply of seeds large enough to make this Maple common in the gardens of this country and of Europe, in which there is every reason to believe that it will flourish.

In the forests of Yezo eight other species of Maple occur. Among them, growing only in the extreme north and on the high mountain-slopes, are a variety of our Mountain Maple, Acer spicatum, so like the New England form of this common tree that it is difficult to distinguish the two plants, and Acer Tataricum, var. Ginnala, a common Manchurian tree, not rare in northern Japan, where it grows in low wet ground, near the borders of streams. This little tree is now well established in American gardens, in which it might be seen more often to advantage, as its flowers are very fragrant, and the leaves of few trees take on more splendid autumnal colors. In Yezo, too, Acer capillipes has been found; this is a species with small racemose flowers, and thin delicate nearly circular lobed leaves, deeply cut on the margins. On Mount Hakkoda, in northern Hondo, where Acer capillipes is extremely abundant at elevations of 2,000 to 3,000 feet above the sea, we found it in October, growing as a stout bush or bushy tree, twelve or fifteen feet in height, with delicate canary-yellow leaves, and secured a supply of ripe seeds.

In Yezo, Acer Japonicum and Acer palmatum are both common; these species, next after Acer pictum, are the most generally distributed Maples in Japan, and the only species which the Japanese cultivate at all commonly. They are both small trees, rarely, if ever, exceeding a height of fifty feet, and both, as is well known, vary remarkably in the size, form, and cutting of their leaves. A few of the varieties of Acer palmatum, particularly the one on which the leaves are divided into narrow lobes, and the one with pendulous branches, are favorites in Japan, where few of the numerous and monstrous forms of this tree, with which we have become familiar of late years in this country, are seen outside of nursery-gardens with foreign connections. Of these two trees the autumn foliage of Acer Japonicum appears the more brilliant; and some individuals of this species which we saw in October, high up on Mount Hakkoda, were as beautiful in color as a good American Scarlet Maple. These two trees have not proved very satisfactory in this country, where they have a way of dying in summer without apparent cause. This is due, perhaps, to the fact that nearly all the plants brought here have been raised from degenerate nursery-stock, obtained in or near the treaty ports; and it will be interesting to watch the behavior here of plants raised from seed gathered in the forests of Yezo. For us these Maples have the advantage of retaining their leaves later in the autumn than our species, which are bare of foliage before the Japanese trees assume their brilliant colors; and this is true of many other Japanese and Chinese plants, like Ampelopsis tricuspidata and Spiræa Thunbergii, for the autumn in eastern Asia is fully a month later than it is in this country.

Acer carpinifolium, which is occasionally seen in our gardens, is evidently extremely rare in Japan. There are a few plants in one of the temple gardens in Nikkō, and I saw a single wild specimen hanging over the bank of a stream in the mountains above Fukushima, on the Nakasendō, and was fortunate in obtaining from it a good supply of seeds. In Nikkō, Acer carpinifolium is a handsome round-topped tree, perhaps thirty feet tall. It is well worth

ACER MIYABEI, Maxm.

growing for its beauty as well as for the unusual form of the leaves, which resemble those of the Hornbeam, for which, at first sight, it might easily be mistaken.

Acer Tschonoskii is common near the margins of Lake Chuzenji in the Nikkō Mountains, and a thousand feet higher is found as a common shrub in the Hemlock forests which cover the slopes rising from Lake Yumoto. It is a small bushy tree, perhaps twenty feet tall, with bright red twigs and ample leaves, not unlike those of Acer capillipes in shape and cutting, although in autumn they turn deep scarlet. We could not find a single seed of this pretty plant, which has probably never been cultivated. Acer rufinerve, hardly distinguishable from the Moosewood of our northern forests (Acer Pennsylvanicum), and Acer cratægifolium, both familiar now in our gardens, are rather common, especially the latter, in all the mountain regions of central Japan, and need no mention here.

Among the rarer and less known species we found Acer diabolicum rather common in the neighborhood of Nikkō, where it is a round-topped tree twenty to thirty feet tall, very like the European Sycamore Maple in habit and general appearance, with dull yellow-green leaves four or five inches across, which apparently do not change color before falling, and large dirty brown fruit covered on the nutlets with fine stinging hairs. This seemed the least beautiful of the Maples which we encountered in the forests of Japan. Acer distylum I only saw in the Botanic Garden in Tōkyō, and Acer pycnanthum, Acer purpurascens, Acer argutum, Acer parvifolium, and Acer Sieboldianum, the last, probably, only a pubescent-leaved variety of Acer Japonicum, I looked for in vain.

Of Maples of the section Negundo, with the male and female flowers on separate plants and pinnate or ternate leaves, there are two species in Japan, — Acer cissifolium and Acer Nikoense. The first is said to be common, and widely distributed from southern Yezo through the mountain ranges of the main island, but I only saw a few small plants in hedge-rows near Nikkō, none of them half the size of specimens which may be seen in some Massachusetts gardens, where Acer cissifolium is a handsome compact round-headed little tree with slender graceful leaves, of a delicate green in summer, and orange and red in late autumn, and where it is one of the most distinct and satisfactory of the Japanese trees which have been tried in our climate.

The second Japanese Negundo (see Plate x.), as it appears in the forests of Japan, is a distinct and beautiful tree, which, if it thrives in this country, will be a real addition to our plantations. Acer Nikoense grows to a height of forty or, perhaps, fifty feet, with a trunk twelve to eighteen inches in diameter covered with smooth dark slightly furrowed bark, and stout rather slender branches, which form a narrow round-topped head. The branchlets are thick and rigid, and are coated at first, like the inner scales of the ovate-acute winter-buds, the young leaf-stalks, the under surface of the young leaflets, the peduncles and pedicels, with short thick pale or rufous villous tomentum; at the end of their first season the branchlets are dark red-brown and are marked with numerous minute lenticular dots. The leaves are ternate, with stout rigid petioles an inch or an inch and a half in length, and ovate or obovate, acute, long-pointed, entire, or remotely and irregularly coarsely and crenately serrate leaflets, the terminal leaflet being long-stalked, symmetrical, and wedge-shaped at the base, the lateral leaflets rounded on the lower, and oblique on the upper edge at the base, and sessile or nearly so. The leaflets are thick and rather rigid, two and a half to five inches long, an inch and a half to two inches broad, conspicuously reticulate venulose, dark yellow-green on the upper surface, pale and coated on the lower surface with pubescence, which is rufous on the stout midribs and broad straight veins; or sometimes they are bright green on the lower surface, and glabrous, except on the midribs and veins. In the autumn the leaflets turn brilliant scarlet on the upper surface, but remain pale on the lower. The flowers are yellow, half an inch across, and nodding, and are borne in short few, usually three-flowered, subsessile terminal corymbs on slender graceful pedicels. The sepals and petals are ovate or obovate, rounded at the apex, and contracted at the base into narrow claws; in the sterile flower, in which the ovary is reduced to a minute rudiment, the stamens, which are inserted between the lobes of the conspicuous disk, are exserted; the filaments are filiform, and the anthers are large and oblong; in the fertile flower the stamens are rudimentary, and not longer than the ovary, which is coated with thick pale tomentum and crowned with a long stout style with revolute stigmas. The fruit is three inches long, with remarkably thick and hard-walled puberulous nutlets, and broad falcate diverging or converging obovate wings rounded at the apex.

Acer Nikoense is not a common, although a widely distributed, species. I saw a number of plants in the temple grounds of Nikkō and on the road between Nikkō and Lake Chuzenji, a single tree near Agematsu on the Nakasendō, and ten or twelve more on the Yusui-toge above Yokokawa. According to Maximowicz,[1] who distinguished this tree nearly thirty years ago, it grows as far south as Nagasaki. Acer Nikoense is practically unknown in gardens, although a well-grown specimen exists in the Veitchian collection in London, and a single small plant was sent from Japan two years ago to the Rixdorf Nurseries in Berlin. A figure of a leaf taken from this plant has been published in a German periodical.

In September we hunted the Nikkō hills in vain for a seed-bearing tree, and had given up all hope of introducing this species. One day late in October, however, we sat down on the rocks in the bed of a torrent far up on the side of Mount Koma-ga-take, in central Japan, to eat our luncheon, when our attention was attracted by some large Maple-seeds which were new to us floating in a pool at our feet. A search on the bank above discovered a single tree of Acer Nikoense, from which the wind was scattering showers of seeds. If we had been a day later, or had selected another resting-place, we should have missed one of the best harvests we made in Japan, as this single tree yielded at least half a bushel of good seeds.

If Acer Nikoense proves hardy and flourishes in our gardens, it will be particularly remarked for the brilliancy of its autumn leaves, which are not surpassed in beauty by those of any other tree which I saw in Japan, and which, unlike those of most trees, are only bright-colored on one surface.

[1] *Mél. Biol.* vi. 370; x. 600; *Bull. Acad. St. Pétersbourg*, t. 76. — Franchet & Savatier, *Enum. Pl. Jap.* i. 90. — Pax, *Bot. Jahrb.* vii. 205; *Gartenflora*, xli. 149. Acer Maximowiczianum, Miquel, *Arch. Néer.* ii. 473, 478.

ACER NIKOENSE, Maxm.

THE SUMACHS AND THE PEA FAMILY.

In eastern North America the small family of the Sabiaceæ has no representative, although Meliosma, which is mostly a tropical and subtropical Asiatic genus, also occurs in Mexico and Central America. In Japan there are three species of this genus, of which only one, Meliosma myriantha, attains the size of a tree. This species grows sometimes to the height of twenty-five or thirty feet, and produces slender trunks and wide-spreading branches; its large thin leaves, which are sometimes eight inches long and three inches broad, of a light delicate green, are its chief attraction as a garden-plant, for the flowers of Meliosma are minute, and the terminal panicles in which they are gathered are loose and long-branched. Only a small portion of the flowers are fertile, so that the fruit, which is a small red berry-like drupe, is sparse and scattered in the clusters, and not at all showy. This plant is new, I believe, in cultivation, and its behavior in our climate will be watched with interest. It can hardly be hoped, however, that Meliosma myriantha will succeed in New England, as in Japan it does not range far north, and in central Hondo, where, although widely distributed, it is not common, it does not rise much above 2,500 feet over the sea-level.

From the Rhus family we miss in Japan the Smoke-tree (Cotinus), a familiar European and western Asiatic type, represented, too, in eastern America by one of the rarest and most local of all our trees. Of the true Rhuses we have in eastern America a dozen species, including three small trees, while in Japan there are five indigenous species, and among them three which can properly be considered trees. The Japanese Lacquer-tree (Rhus vernicifera), which has played a conspicuous part in the development of the mechanical arts of China and Japan, and which is certainly the most valuable plant of the genus to man, is not a native of Japan, where it was carried long ago from China, and although much cultivated, especially in northern Hondo, I saw no indications that it is growing spontaneously or anywhere establishing itself in the forest.

The Japanese Rhuses are not as ornamental in the autumn as our Sumachs, as none of them bear fruit covered with the long red hairs which give to the fruit-clusters of the American plants their dense appearance and brilliant color; but the flowers of the Asiatic Rhus semialata, a common small tree distributed from the Himalayas to Japan, which are white, and produced in large terminal panicles, are much more beautiful than the yellow-green flowers of any of our Sumachs, and in August and September, when this tree blossoms in Japan, it is a striking object in the shrubby coppice-growth which so often covers the low mountain-slopes. In autumn Rhus semialata is one of the most brilliantly colored plants of the Japanese forest; and very few Japanese plants succeed so well in our climate. It is from a gall formed on the leaf of this tree that the dye with which married women in Japan discolor their teeth, as a sign of domestic bondage, is obtained.

Economically a more important tree, as from it the Japanese obtained their principal supply of artificial light before the introduction of American and Russian petroleum, Rhus succedanea is less interesting in flower, at least, than Rhus semialata; it is a southern species, still much cultivated on the southern islands, and in Tōkyō seen only in gardens. In habit, although it grows to a larger size, it much resembles our Stag-horn Sumach; the leaflets are narrower, the flowers are produced in slender few-flowered clusters pendulous in fruit; and the drupes covered with a thick coat of the pale waxy exudation, to which this species owes its name and value, are much larger. Rhus succedanea will, no doubt, flourish in the southern states, and it is not improbable that it will prove hardy as far north as Philadelphia; it will certainly never be grown, however, in the United States for the wax it might be made to yield, and as an ornamental plant, while it is, of course, interesting, it is inferior to the American Sumachs.

Rhus trichocarpa, which, so far as I know, is not in our gardens, should be cultivated for the extraordinary beauty and brilliancy of the leaves in autumn, when they assume the brightest scarlet and orange tints. It is a slender tree, sometimes twenty or twenty-five feet high, and very common in the forests of Yezo and on the mountains of central Hondo. The leaves are eighteen to twenty inches long, with dark red puberulous midribs and broadly ovate long-pointed short-stalked membranaceous leaflets, slender panicles of flowers, which open in July, and pendulous fruit-clusters, with large pale prickly drupes ripening in August or early in September. Neither the flowers nor the fruit are attractive, and there is nothing very distinct in the appearance of this tree, except in the autumn, when, however, it is so beautiful that if it succeeds here I believe it will prove one of the best introductions of recent years.

Of the poisonous species of Rhus I did not see the pinnate-leaved Rhus sylvestris, which is said to be a small shrub and a native of the southern part of the empire; on the Hakone Mountains, where it is reported to grow, I looked for it in vain; but our Poison Ivy is one of the common plants in all the central parts of Hondo and in Yezo, where it grows to its largest size and climbs into the tops of the tallest trees. The leaves of the Japanese plant are larger than they usually grow on the American form; they are thicker, too, and more leathery, and turn to even more brilliant autumn colors, often to deep shades of crimson, which are rarely seen on this plant in America. In October no other vine is so handsome in Japan.

Japan is remarkably poor in arborescent Leguminosæ, with only three species in three genera, while here in eastern America there are twenty species in a dozen genera. The best known Japanese tree of the family is Albizzia Julibrissin, a small Mimosa-like tree which grows from Persia to Japan, and through cultivation has become naturalized in our southern Atlantic states and in most other warm temperate countries. Familiar now in this country is Maackia Amurensis, which, introduced many years ago from the valley of the Amour, is now sometimes cultivated in northern gardens. This little tree, which, under favorable conditions, rises occasionally to the height of thirty or forty feet, is common in all the forest regions of northern Japan, and is not rare on the mountains of central Hondo. The Japanese form produces larger and more numerous flower-spikes and larger fruit than the mainland tree, as we see it in this country; and it is not improbable that it will prove a more desirable garden-plant. In Yezo, the wood, which is hard, close-grained, and pale brown in color, is manufactured into many small objects of domestic use, and is considered valuable.

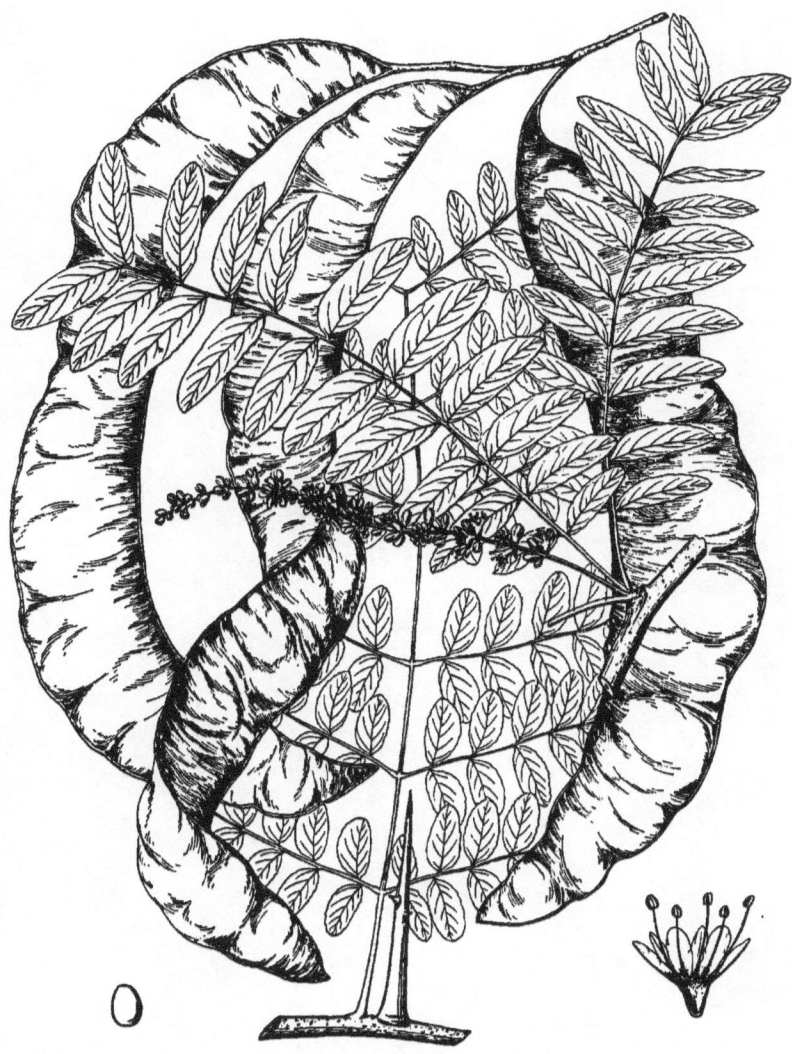

GLEDITSIA JAPONICA, Miq.

The third Japanese leguminous tree is Gleditsia Japonica,[1] which, in most essential characters, resembles our North American Gleditsia triacanthos, but the leaflets are broader and more lustrous, and the bark, instead of being dark brown, is quite pale. Although it does not grow to the great size of the American species, the Japanese Gleditsia is, perhaps, a more beautiful tree. It is (see Plate xi.) a tree sixty to seventy feet in height, with a trunk occasionally three feet in diameter, and stout branches horribly armed with flattened, often branched lustrous red-brown spines, two or three inches in length. The branchlets are remarkably stout as compared with those of our species, and are covered with bright green bark, marked with orange-colored elevated lenticular spots. The leaves are ten to twelve inches long, with broad ovate-acute remote leaflets, or they are sometimes bipinnate as on our species, with smaller leaflets. The male flowers (the female inflorescence I have not seen) are very similar to those of Gleditsia triacanthos, although they are rather larger and the racemes are longer and less closely flowered. The pods are compressed and thin-valved, like those of our northern tree, ten or twelve inches long and an inch and a half broad, but the seeds, instead of being placed close to the ventral suture of the pod, are sometimes nearer the middle and surrounded by the pulp, which is more abundant in the Japanese than in the American species. This pulp is used by the Japanese in washing cloth, and long strings of the pods are displayed for sale in many towns of northern Japan, where Gleditsia Japonica grows, not very abundantly, according to my observations, near the banks of streams at the sea-level. It is common, and reaches its largest size on the banks of the Kisogawa and other streams of central Japan, at an elevation of some two thousand feet. Here it grows sometimes close to the water's edge, in rich humid soil, or is as often found at a considerable distance above the water, growing on dry gravelly slopes. By Rein this tree is said to be often planted in the neighborhood of villages in Japan, but I saw no specimens, except in the scientific gardens of Tōkyō, which did not have the appearance of growing naturally.

As an ornamental tree, Gleditsia Japonica, as it appeared on the mountains of Japan, is a more beautiful tree than any of the species common in cultivation, and it may be expected to become a valuable addition to the list of exotic trees suitable for the decoration of the parks and avenues of the United States and Europe.

[1] Gleditsia Japonica, Miquel, *Prol. Fl. Jap.* 242. — Franchet & Savatier, *Enum. Pl. Jap.* i. 114 ; ii. 325. — Maximowicz, *Mél. Biol.* xii. 452.

THE ROSE FAMILY.

TREES of the Rose family in the flora of Japan are not numerous as compared with that of eastern America, and among them there is not one of first-rate value as a timber-tree. Horticulturally they are more important, and Japanese gardens owe much of their interest to species of Prunus. Although the most popular garden-tree in Japan, Prunus Mume is probably not Japanese at all, but a native of Corea, where Mr. Veitch found it planted as a shade-tree along the borders of the highroads. This is the tree which all foreign writers upon Japan speak of as the Plum, although it is really an Apricot. In cultivation Prunus Mume produces white, rose-colored, or red, and often double flowers, which appear before the leaves in February and March, and are revered as harbingers of spring. The Mume is planted in nearly every Japanese garden of any pretensions, and is one of the most universally used pot-plants. Care and labor are often expended in producing dwarfed, contorted, or pendulous-branched specimens, which sometimes command what seem exorbitant prices.

A more interesting tree than Prunus Mume is the Japanese Cherry, Prunus Pseudo-Cerasus, the largest tree of the Rose family in the empire, and, next to the Apricot, more cultivated for flowers by the Japanese than any other tree. In the forests of Yezo, Prunus Pseudo-Cerasus occasionally rises to the height of eighty feet, and forms a trunk three feet in diameter. In the character of the bark, in habit and general appearance, it much resembles the European Cherry, the wild type of the familiar Cherry-tree of our gardens and orchards, and as it appears in the forest it might well be mistaken for that species. The Japanese Cherry is common in Yezo and in all the mountain regions of Hondo up to 5,000 or 6,000 feet above the sea-level, and often forms a considerable portion of the forest-growth, although, in Hondo, all large trees appear to have been cut. In the early autumn it is conspicuous in the landscape and very beautiful, as the leaves turn deep scarlet and light up the forest before the Maples assume their brightest colors. For centuries the Japanese have planted these Cherry-trees in all gardens and temple grounds, and often by the borders of highways, as at Mukōjima, near Tōkyō, where there is an avenue of them more than a mile in length along the banks of the Sumi-da-gawa, and at Koganei, where, a century and a half ago, 10,000 Cherries were planted in an avenue several miles long. The flowering of the Cherry-tree is an excuse for a holiday, and thousands of men, women, and children pass the day under these long avenues in more or less hilarious contemplation of the sheets of bloom. The flowers of the wild tree are single, white, and of the size of those of the garden Cherry, but, not unnaturally, many varieties have been produced during the centuries it has been a garden-plant. Bright red and pink single-flowered varieties are common in Japan, as well as many double-flowered forms. Of these several have been introduced into this country and Europe, and are now well known in our gardens, where, however, they do not flower as freely as they are represented to flower in their native land. Prunus Pseudo-Cerasus is a plant of cold climates, and

PRUNUS MAXIMOWICZII, Rupr.

great summer heat evidently does not suit it, as in Tōkyō the trees never grow to a great size, and by midsummer are leafless; so that, except during the short blooming season, the excessive use of this tree is a real injury to the appearance of the gardens and promenades of the capital and of other southern cities.

Prunus Pseudo-Cerasus is of some value as a timber-tree, producing hard, close-grained red wood, which is hardly to be distinguished from that of the European Cherry. It is used in considerable quantities for all sorts of wooden dishes and other small articles of domestic use. Rather curiously, perhaps, no attention has been paid to improving the size and quality of the fruit, which is not larger than a small pea, with a thin layer of flesh.

The pendulous-branched Cherry-tree, with precocious pink flowers, now common in our gardens, where it is known as Prunus pendula, is often cultivated by the Japanese, who, however, do not appear to feel the same regard for this graceful tree that the Apricot and the Cherry inspire. I never saw it growing wild, and cannot refer the cultivated plants to a wild type. Specimens fifty or sixty feet high, with wide-spreading fountain-like heads, are not uncommon in old temple gardens in many of the cities of Hondo. This beautiful tree thrives perfectly in our climate, and in early spring, when its branches are covered with its pale pink pendulous flowers, no tree is more beautiful.

I did not see the Cherries in bloom, and most of them had dropped their fruit before I reached Japan; several species described by botanists I did not see at all, and of several others I obtained a very superficial idea; and there is evidently still much to be learned of the proper limitation of described east Asian species and varieties, and of their geographical distribution. Among the little known species, Prunus Maximowiczii (see Plate xii., made from material for which I am indebted to Professor Miyabe), seems to deserve the attention of horticulturists.

As I saw it in Yezo, Prunus Maximowiczii[1] is a tree twenty-five to thirty feet in height, with a slender trunk and branches covered with smooth pale or light red bark. The young branchlets and petioles, the under surface of the unfolding leaves, and the branches of the inflorescence are coated with rusty pubescence which only partly disappears during the season. The leaves are elliptical or elliptical-obovate, contracted at the apex into long slender points, wedge-shaped or rounded at the base, long-petiolate, coarsely and doubly serrate, thin, light green on the upper, and paler or rufous on the lower surface. The stipules are foliaceous, lanceolate-acute, coarsely serrate, an inch long, or rather shorter than the petioles, and deciduous. The flowers, which, in the neighborhood of Sapporo, appear in May, are produced on long slender pedicels in axillary racemes three or four inches long, and conspicuous from their large foliaceous bracts, which are coarsely serrate with gland-tipped teeth; they are half an inch across when expanded, with leafy serrate hairy calyx-lobes and obovate or orbicular white petals. The fruit ripens in July, and is oblong and rather less than a quarter of an inch long. In Japan, Prunus Maximowiczii does not appear to be a common tree. I saw a few specimens on the hills near Sapporo and a single tree on the main island, where it is said

---

[1] Prunus Maximowiczii, Ruprecht, *Bull. Phys. Math. Acad. St. Pétersbourg*, xv. 131. — Maximowicz, *Fl. Amur.* 80; *Mél. Biol.* xi. 700. — F. Schmidt, *Reisen in Amurland*, 125. — Franchet & Savatier, *Enum. Pl. Jap.* i. 118. — Forbes & Hemsley, *Jour. Linn. Soc.* xxiii. 219.

by Maximowicz to grow in several of the mountain provinces; it also inhabits Saghalin, Corea, and eastern Manchuria, where it was discovered.

The common Prunus Padus of Europe and northern Asia reaches northern and central Yezo, where it is not rare in low ground in the neighborhood of streams, and where it grows to a considerable size. A much more common tree in Yezo and in the elevated forests of Hondo is Prunus Ssiori, another Bird Cherry, always easily distinguished by its pale, nearly white bark. It is a handsome glabrous tree with oblong membranaceous leaves and long graceful racemes of small flowers, and is well worth introducing into our plantations as an ornamental plant. It grows also in Saghalin, where it was discovered by Schmidt, in Manchuria, and in western China. The wood of Prunus Ssiori is very hard and close-grained, and is used by the Ainos for numerous domestic purposes.

Prunus Grayana, the third Japanese Bird Cherry, is common in all the mountain forests of Hondo, and extends across the straits of Tsugaru into southern Yezo. It is a small tree, twenty to thirty feet high, with a slender trunk, ample membranaceous long-pointed setaceo-serrate leaves, biglandular at the base, but without glands on the petioles, a peculiarity which best distinguishes this species from Prunus Padus, although the hair-like teeth of the leaves are characteristic and apparently constant.

Of true Plums there are in the flora of eastern America no less than nine or ten indigenous species, of which six are considered trees; in some parts of the country these plants are exceedingly common, and in early spring enliven forest glades, or the seacoast of the north with their profuse and fragrant flowers; but Japan apparently possesses no indigenous Plum-tree, and although Plums are sometimes cultivated in the neighborhood of Japanese houses, they are by no means common, and the fruit which is offered for sale in the markets is not abundant or of good quality.

In recent years a good deal has been heard in this country of Japanese Plums, which are now successfully cultivated in the southern states. Some of these varieties have possibly been made in Japanese gardens, but the original stock from which they have all been derived is probably some south China or Indian species of doubtful identity — perhaps, as has been suggested, the Prunus triflora of Roxburgh, an obscure plant, which is possibly a form of Prunus domestica. But the parentage of the so-called Japanese Plums will not be satisfactorily settled until competent botanists have explored western and southwestern China, where must be solved many of the problems which relate to the origin and geographical distribution of many cultivated plants.

The Japanese form of the Old World Mountain Ash, Pyrus aucuparia, is a common tree in Yezo and on all the high mountain ranges of Hondo, where it is sometimes twenty or thirty feet in height, and always conspicuous in autumn from the brilliant orange and scarlet colors of the foliage. This peculiarity will give to this tree a horticultural value, although, except in its more glabrous buds, it does not vary in any marked way from the Mountain Ash of Europe and northern Asia.

Pyrus sambucifolia, which is much like the American plant, although described as a small shrub, is said to be abundant in northern and eastern Yezo and on the Kurile Islands. I only saw it in the Botanic Garden at Sapporo, where Professor Miyabe has established a remarkable

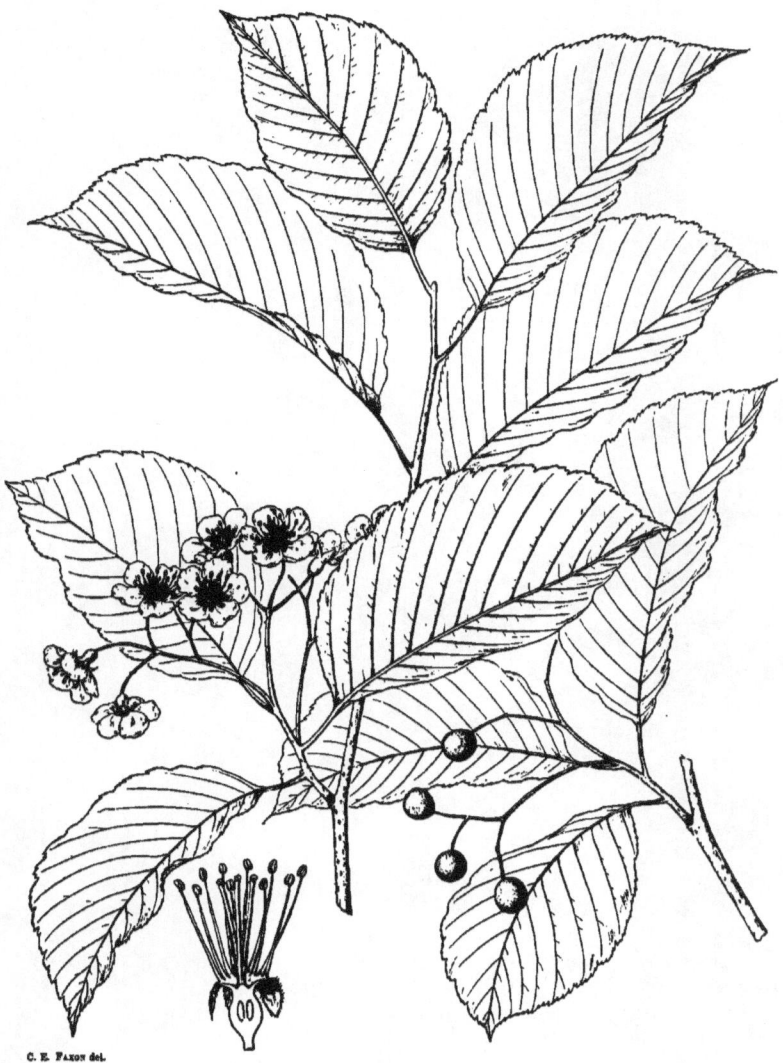

PYRUS MIYABEI, Sarg.

collection of Hokkaido plants. The third Japanese Mountain Ash, Pyrus gracilis, is, I believe, unknown in cultivation; it is a particularly well-marked species, with woolly buds, leaves only four or five inches long, oval or oblong leaflets rounded or acute at the apex, and pale on the lower surface, orbicular incisely serrate stipules an inch or more across, minute flowers in small few-flowered clusters, and oblong fruit barely an eighth of an inch long. Pyrus gracilis inhabits mountain forests in Kyūshū and in central Hondo, where, however, I did not succeed in finding it.

Aria is represented in Japan by two handsome trees, the first, Pyrus lanata [1] (I follow Hooker in the *Flora of British India* in referring the Japanese plant to the Pyrus lanata of Don, which grows also on the Himalayas from Cashmere to Kumaon), is not rare in central Japan, where it is principally found at about 5,000 feet elevation above the sea-level on the lower edge of the great Hemlock forest. Here it is a tree thirty or forty feet in height, with a trunk six to eight inches in diameter, slender light red branchlets marked with white dots, and oblong obtuse winter-buds covered with pale chestnut-colored imbricated scales. The leaves are three or four inches long, two or three inches wide, broadly oblong to ovate-lanceolate, acute at the apex, slightly lobulate and serrate, dark green and mostly glabrous on the upper surface, and silver-white and more or less thickly coated with tomentum on the lower surface. The flowers I have not seen, but the fruit is subglobose to oblong, one third of an inch long, bright scarlet, and marked with pale lenticels. The second Japanese species of Aria (see Plate xiii.) is a tree fifty or sixty feet in height, with a trunk covered with pale smooth bark, and occasionally a foot and a half in diameter, slender branches, which form a narrow oblong head, and red branchlets marked by oblong lenticular dots. The leaves are ovate, acute, often long-pointed at the apex, rounded or sometimes wedge-shaped at the base, serrate with incurved teeth, or often coarsely and doubly serrate above the middle, thin, or subcoriaceous at maturity, dark green on the upper surface, pale on the lower, two or three inches long and one or two inches broad, with thick prominent midribs, straight parallel veins, and slender petioles one or two inches in length. The flowers, which appear near Sapporo early in June, are borne in loose spreading long-branched few-flowered corymbs, and are half an inch in diameter. The calyx-lobes are ovate, acute, densely coated on the inner surface with thick white tomentum, and much shorter than the oblong white petals, rounded at the apex and contracted at the base into short claws more or less covered with tufts of long white hairs. The stamens are exserted, with filiform filaments enlarged at the base, and rather longer than the two spreading styles. The fruit ripens in September, and is oblong or subglobose, the size of a pea, light red, and conspicuously marked by the scar left by the deciduous calyx. Unfortunately the name Pyrus alnifolia, which has been given to this tree, is not applicable, it having been previously applied by Sprengel, in 1825, to an entirely different plant, Amelanchier alnifolia, and as a new name must be found for it, I am glad of the opportunity of associating with this fine tree that of Professor Kingo Miyabe, whose

---

[1] Pyrus lanata, Don, *Prodr. Fl. Nepal.* 237. — Hooker f. *Fl. Brit. Ind.* ii. 375.

Sorbus Aria, var. Kamaonensis, Maximowicz, *Mél. Biol.* ix. 173.

Sorbus lanata, Wenzig, *Linnæa*, xxxviii. 61.

knowledge of the flora of Hokkaido is unrivaled. Pyrus Miyabei[1] is one of the common trees of the forests of central Yezo, and, according to Maximowicz, it inhabits the province of Nambu, in Hondo, and southern Manchuria. So far as I know, it has not been introduced into our gardens, where it may be expected to flourish.

Of true apple-trees, there is apparently only a single indigenous species in Japan, the Pyrus Toringo of Siebold. This is the tree which is often cultivated in American and European gardens as Pyrus Malus floribunda, Pyrus microcarpa, Pyrus Parkmani, Pyrus Halleana, Pyrus Sieboldii, and Pyrus Ringo. It is a common and widely distributed plant in Japan, growing from the sea-level in Yezo to elevations of several thousand feet in central Hondo, usually in moist ground in the neighborhood of streams. Sometimes it is a low bush, but more often a tree fifteen to thirty feet in height, with a short stout trunk and spreading branches. The leaves are exceedingly variable, and on the same plant are often oblong, rounded or acute at the apex, or broadly ovate or more or less deeply three-lobed. The fruit, which, like that of the Siberian Pyrus baccata, loses the calyx before it is fully ripe, resembles a pea in size and shape, and in color varies from bright scarlet to yellow. In early spring Pyrus Toringo is one of the most beautiful of the trees found in our gardens, where it is perfectly hardy, and where it covers itself every year with fragrant pink or red single or semi-double flowers.

Pyrus Sinensis, the common cultivated Pear-tree of Japan, although now growing spontaneously in some mountain regions, is probably a native of northern China and Manchuria; and the only indigenous Pear-tree is Pyrus Tschonoskii. This interesting and handsome tree was first described by Maximowicz,[2] whose collector, Tschonoski, brought to him from the slopes of Fuji-san a single fruit and a portion of a leaf, now preserved in the herbarium of the Imperial Botanic Garden at St. Petersburg. Nothing more was seen of it until Mr. J. H. Veitch and I encountered in the woods near Nikkō a single tree of a Pyrus, which by subsequent comparison with Tschonoski's specimen, proved to have been this tree. It is evidently rare, for I only saw it in two other localities, — in the grounds of a temple near Nokatsu-gawa, where there was a single specimen, and in the woods at the head of the Ysuitoge, near Karuizawa, at the base of the volcano Asama-yawa, in central Hondo, where there were two or three trees. Pyrus Tschonoskii (see Plate xiv.), which is a Pear-tree, rather than an Apple as described by Maximowicz, is, as we saw it, a tree thirty to forty feet in height, with a trunk about a foot in diameter, covered with smooth pale bark, and a narrow round-topped head. The branchlets are stout and terete, and are marked with small oblong or circular orange-colored lenticels; during their first summer they are red-brown, rather lustrous, covered with loose pale tomentum, and encircled at the base by the conspicuous ring-like scars left by the falling of the inner scales of the winter-buds; later they grow darker, and sometimes nearly black. The winter-buds are ovate, obtuse, and rather less than a quarter of an inch long, and are covered with loosely imbricated chestnut-brown lustrous scales, tomentose above the middle, and ciliate on the margins. The leaves are ovate, acumi-

---

[1] Pyrus Miyabei.
Cratægus alnifolia, Siebold & Zuccarini, *Abbild. Acad. Münch.* iv. 130. — Regel, *Act. Hort. Petrop.* i. 125.
Sorbus alnifolia, Miquel, *Ann. Mus. Lugd. Bat.* i. 249. —
Maximowicz, *Mél. Biol.* ix. 173. — Wenzig, *Linnæa*, xxxiii. 61.
Aronia alnifolia, Decaisne, *Nouv. Arch. Mus.* x. 100.

[2] *Mél. Biol.* xii. 105.

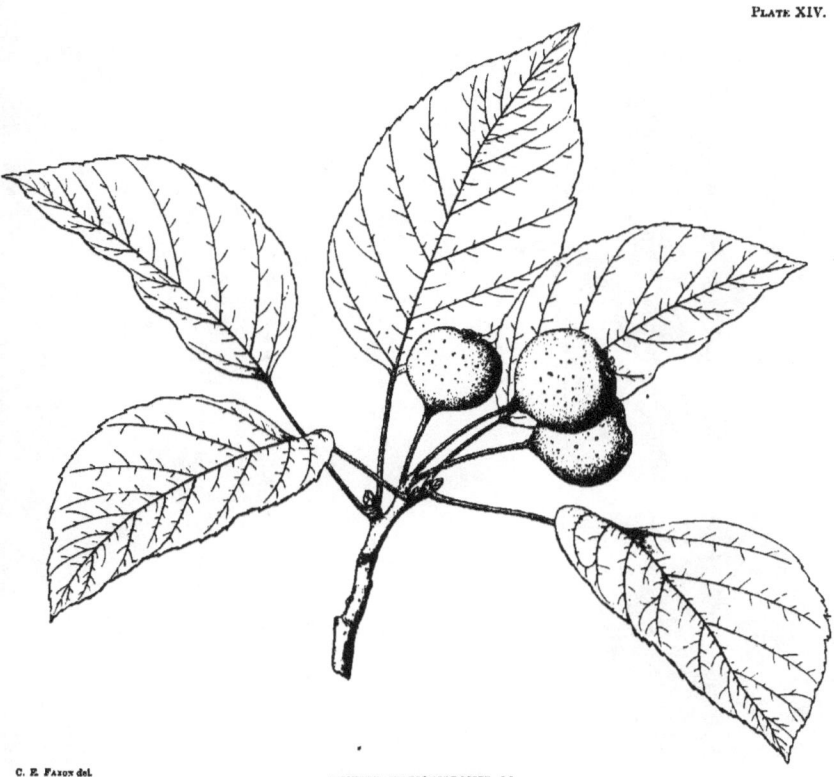

PYRUS TSCHONOSKII, Maxim.

nate, unequally rounded or wedge-shaped at the base, and coarsely and unequally serrate with rigid glandular teeth, which are largest and most unequal above the middle of the leaf; they are thick and firm, dark green, lustrous and pilose on the upper surface, coated on the lower surface and on the petioles with thick pale, close tomentum, four to five inches long and two to three inches broad, with stout midribs impressed on the upper side, and five or six pairs of conspicuous oblique veins running to the principal teeth and connected by reticulate cross veinlets; they are borne on slender terete petioles an inch and a half in length. The flowers are unknown. The fruit, which is usually solitary, or is sometimes in clusters of two or three, is obovate, pointed at the base, and crowned with the thickened and partly immersed calyx-lobes, which are triangular, obtuse, and covered with a thick coat of dense white tomentum; it is an inch long, two thirds of an inch broad, of a dull yellow color, and rosy-red on one side, with a thick skin covered with pale lenticels, and austere coarse granular flesh. The seed is a quarter of an inch long, obliquely obovate, acute at the base, and covered with a light red-brown shining coat. The fruit is borne on a stout rigid stem an inch to an inch and a half long and coated with pale loose tomentum, especially toward the much thickened apex.

Cratægus, which, in eastern America, abounds with many species which are conspicuous features of vegetation in different parts of the country, is only represented in Japan by Cratægus chlorosarca, one of the black-fruited group related to Cratægus Douglasii of our Pacific states, which it much resembles. It is not rare in the neighborhood of Sapporo, where it grows near streams in low wet soil, and apparently does not range south of Yezo. The flowers are not large, and as a garden-plant this species has little to recommend it.

The Saxifrage family, which is conspicuous in Japan with a large number of shrubs, including some which have become important features in our gardens, has only a single Japanese arborescent representative; this is the now well-known Hydrangea paniculata, which is one of the most common northern and mountain plants, and which occasionally in favorable situations, especially on the hills of central Yezo, becomes a tree twenty-five to thirty feet in height, with a short well-formed trunk a few inches in diameter and branches stout enough for a man to climb into. From the branches the Ainos make their pipes.

THE WITCH-HAZEL AND ARALIA FAMILIES.

In the Witch-hazel family, Distylium racemosum, an evergreen tree of the southern islands and of southern China, with peculiar and exceedingly hard dark-colored valuable wood, will require in this country the mild climate of the extreme southern states and of California. The Japanese Hamamelis, however, is already an inhabitant of our gardens, where, unlike the American species which flowers in the autumn, it produces its orange or wine-colored flowers in March. Hamamelis Japonica is one of the common forest-shrubs or small trees in its native country, where specimens occasionally occur thirty or forty feet in height, with stout straight trunks and broad shapely heads. In the autumn the leaves turn bright clear yellow; but on one form which we found on Mount Hakkoda, near Aomori, with small thick often rounded leaves (Hamamelis arborescens of Hort., Veitch), they were conspicuous from their deep rich vinous red color. This may, perhaps, prove to be a second Japanese species.

We were fortunate in securing a good supply of ripe seeds of the little known Disanthus cercidifolia of Maximowicz, a curious and interesting member of the Witch-hazel family (see Plate xv.). Disanthus, of which only one species is known, is a shrub with slender spreading branches, eight or ten feet high, stout terete red-brown branchlets conspicuously marked with pale lenticels, and obtuse buds covered with chestnut-brown imbricated scales. The leaves are suborbicular, rounded and minutely mucronate at the apex, or rarely orbicular-ovate and sharp-pointed, cordate or rarely truncate at the base, entire, palmately five or seven nerved, dark blue-green on the upper surface, pale on the lower, thick and firm, or ultimately subcoriaceous, and three or four inches long and broad, with reticulated veinlets and stout petioles one or two inches long and thickened at the base. In the autumn they turn deep vinous red or red and orange. The flowers appear in October, when the fruit developed from the flowers of the previous year ripens; they are dark purple, sessile, base to base, in two-flowered heads on slender-ridged peduncles produced from scaly buds, and are each surrounded by three thick ovate obtuse woolly closely imbricated bracts which form the apparent connective between the two flowers. The calyx is five-parted; the divisions which are imbricated in æstivation are ovate, obtuse, latitudinally unequal, reflexed, and much shorter than the five lanceolate acute petals imbricated in æstivation, spreading into a star-shaped corolla, and slightly incurved at the apex. The stamens are as long as the lobes of the calyx and are inserted on its base opposite the petals; the filaments are short and broad, and as long as the anthers, which are nearly as broad as long, attached on the back, two-celled, and extrorse, the cells opening longitudinally. The ovary is superior, ovate, compressed, two-celled, gradually contracted into two short spreading styles stigmatic at the apex; the ovules are numerous in each cell, suspended from its apex, and anatropous. The fruit is a woody ovoid two-celled capsule, which opens loculicidally, with a thin cartilaginous inner coat separable from the thick hard outer covering. The seed, of which there are a number in each cell, is ovate,

DISANTHUS CERCIDIFOLIA, Maxm.

acute, compressed, angled by mutual pressure, with a thick hard dark chestnut-brown lustrous coat, an oblong pale lateral hilum, and thin albumen surrounding the terete embryo, with its long erect radicle and thick ovate cotyledons.

Disanthus cercidifolia[1] is not rare in the valley of the Kisogawa on the Nakasendō, in central Hondo, where it is occasionally found, covering steep hillsides with thickets sometimes a quarter of an acre in extent. In habit and in the autumn color of its leaves Disanthus is one of the most beautiful shrubs which I saw in Japan, and if it flourishes in our gardens it should prove one of the best plants of its class recently introduced into cultivation.

The Aralia family has no representative in eastern North America outside of the genus Aralia and only one woody plant, Aralia spinosa, a small tree of the middle and southern states. In Japan the family appears in no less than eight genera. The Ivy of Europe reaches Japan, where it is rather common in the south, although we did not meet with it north of the Hakone Mountains and the region about Fuji-san. Helwingia, a genus with two species of shrubs, remarkable in this family for the position of the flowers which are produced on the upper surface of the midribs of the leaves, is Japanese and Himalayan. In Japan Helwingia ranges to southern Yezo, where, in the peninsula south of Volcano Bay, in common with a number of other plants, it finds its most northern home.

In the flora of Japan, Fatsia is represented by the handsome evergreen plant, Fatsia (Aralia) Japonica, now well known in our conservatories, an inhabitant of the extreme southern part of the empire, although often cultivated in the gardens of Tōkyō, both in the open ground and in pots; and by Fatsia horrida, a low shrub with stout well-armed stems, large palmately lobed leaves and bright red fruit, which is also common on the mountains of the northwest coast region of North America, from Oregon to Sitka. In Japan we found it growing under the dense shade of the Hemlock forests on steep rocky slopes above Lake Yumoto, in the Nikkō Mountains, at an elevation of 5,000 feet above the sea-level, and in Yezo. The third member of the genus, Fatsia papyrifera, from the thick pith of whose branches the Chinese rice-paper is made, and an inhabitant of central and southern China and of Formosa, is frequently seen in Tōkyō gardens, as it is in those of the United States and Europe. In Yezo is found a representative of the Manchurian and Chinese genus Eleutherococcus, a shrub still to be introduced into our gardens, and Panax repens, a delicate herb with trailing stems and bright red fruit, which manages to live on mountain-slopes under the dense shade of Bamboos, while Dendropanax, a tropical genus of trees and shrubs of the New World, as well as of the Old, reaches southern Japan with a single shrubby species, Dendropanax Japonicum.

Aralia is more multiplied in species in eastern America, where six are known, than it is in Japan, whose flora contains only two, although a third, the Ginseng (Aralia quinquefolia), a native of Manchuria, northern China, and the United States, has been cultivated for centuries in Japan for the roots, which the Chinese esteem for medicine and buy in large quantities, sometimes paying fabulous prices for them, especially for the wild Manchurian roots, which are considered more valuable than those obtained from North America or from plants cultivated in Japan, or in Corea, where Ginseng cultivation is one of the most important branches

[1] Maximowicz, Mél. Biol. vi. 21 (1866).

of agriculture. Curiously enough, this North American and Chinese species was first made known to the outside world by Kaempfer's description of the plants cultivated in Japan.

Of the indigenous Aralias of Japan, Aralia cordata is an herb with large pinnate leaves and long compound racemose panicles of white flowers, which are followed by showy black fruits. In habit and general appearance it resembles our North American Spikenard, Aralia racemosa; but it is a larger and handsomer plant, and well worth a place in the wild garden. In Japan Aralia cordata is often cultivated in the neighborhood of houses for the young shoots, which, as well as the roots, are cooked and eaten. The second Japanese Aralia only differs from our American Aralia spinosa in its rather broader and more coarsely serrate leaflets, and in the character and amount of pubescence which covers their lower surface. Aralia spinosa, var. elata, is a common tree in Yezo and in all the low mountain region of northern and central Hondo. It usually selects rather moist soil, and sometimes, under favorable conditions, rises to the height of thirty or forty feet and forms a straight well-developed trunk. In Hondo large plants are rare, probably owing to the fact that the forests on the low and accessible mountain-slopes are frequently cut off, but the shrubby covering of such hills is almost always brightened in September by the great compound clusters of the white flowers of the Aralia which rise above it. The Japanese form does not appear to be much known in gardens, although young plants have lately been raised in the Arnold Arboretum from seed sent from Japan a few years ago by Dr. Mayr; and it is the Manchurian variety, known as Aralia Chinensis, or as Dimorphanthus Manchuricus, that is usually seen in our gardens, from which the American form, the type of the species, appears to have pretty nearly disappeared, although the name is common enough in nurserymen's catalogues.

But of all the Araliaceæ of Japan, Acanthopanax is the most interesting to the student of trees. It is a small genus of about eight species of trees and shrubs, all members of tropical Asia, and of China and Japan, where half a dozen of them have been found. The most important of the Japanese species are Acanthopanax ricinifolium and Acanthopanax sciadophylloides. Of the other species, Acanthopanax innovans is a small tree, of which I saw young plants only on the Nakasendō, without flowers or fruit, and which is still to be introduced. Acanthopanax aculeatum, a shrub or small tree, with lustrous three or five parted leaves, is much planted in Japan in hedges, and is hardy in southern Yezo, where, however, it has been introduced. Acanthopanax trichodon, of Franchet & Savatier, a doubtful species, which, from the description, must closely resemble Acanthopanax aculeatum, we did not see; but Acanthopanax sessiliflorum of Manchuria and northern China, and an old inhabitant of the Arnold Arboretum, we found evidently indigenous near Lake Yumoto, in the Nikkō Mountains, on the Nakasendō and in Yezo.

Acanthopanax sciadophylloides is still unknown in our gardens, and we were fortunate in securing an abundant supply of seeds. It is a handsome shapely tree sometimes forty feet in height, with a trunk a foot in diameter, covered with pale smooth bark, short small branches which form a narrow oblong round-topped head, and slender glabrous unarmed branchlets. The leaves are alternate, and are borne on slender petioles with enlarged clasping bases and four to seven inches in length, and are composed of five, or rarely of three, leaflets; these are oval or obovate, long-pointed, wedge-shaped at the base, coarsely serrate

PLATE XVI.

ACANTHOPANAX RICINIFOLIUM, Seem.

with incurved teeth tipped with long slender mucros, membranaceous, dark green on the upper surface, and pale or sometimes almost white on the lower surface, quite glabrous at maturity, five or six inches long and two or three inches broad, with stout petiolules sometimes an inch in length, broad pale midribs, and about seven pairs of straight primary veins connected by conspicuous transverse reticulate veinlets. The flowers appear in early summer, on slender pedicels in few-flowered umbels arranged in terminal panicles five or six inches across, with slender branches, the lower radiating at right angles to the stem, the upper erect. According to Franchet & Savatier, the flowers are five-parted, with acute calyx-teeth, oblong-obtuse greenish white petals, and two united styles. The fruit, which is the size of a pea, is dark blue-black, somewhat flattened or angled, crowded with the remnants of the style, and contains two cartilaginous flattened one-seeded stones. This handsome species inhabits the mountain forests of Nikkō, where it is not common. Later we found it in great abundance on Mount Hakkoda, in northern Hondo, and in central Yezo, where it is common in the deciduous forests which clothe the hillsides. Here it apparently attains its largest size, and grows with another species of this genus, Acanthopanax ricinifolium, the largest Aralia of Japan. I have followed the Japanese botanists in referring this tree to the Panax ricinifolia of Siebold & Zuccarini, although the plant cultivated in our gardens and in Europe as Acanthopanax ricinifolium or Aralia Maximowiczii is distinct from the Yezo tree in the more deeply lobed leaves with much broader sinuses between the lobes. A single individual similar to the plant of our gardens I saw growing in the forest near Fukushima, in central Japan, but, unfortunately, it was without flowers or fruit. And as I was unable to find any leaves on the Yezo trees with the broad sinuses of this plant or any intermediate forms, it will not be surprising if the forests of Japan are found to contain two species of simple-leaved arborescent Acanthopanax, in which case it will be necessary to examine Siebold's specimens to determine which species he called Panax ricinifolia.

In the forests of Yezo, where it is exceedingly common, Acanthopanax ricinifolium (see Plate xvi.), as it will be called for the present at least, is a tree sometimes eighty feet in height, with a tall straight trunk four or five feet in diameter, covered with very thick dark deeply furrowed bark, and immense limbs which stand out from the trunk at right angles, like those of an old pasture Oak, and thick reddish-brown mostly erect branchlets armed with stout straight orange-colored prickles with much enlarged bases. The leaves are nearly orbicular, although rather broader than long, truncate at the base, and divided to a third of their width or less by acute sinuses into five nearly triangular or ovate acute long-pointed lobes finely serrate with recurved callous-tipped teeth; they are five to seven ribbed, seven to ten inches across, dark green and very lustrous on the upper surface, and light green on the lower surface, which is covered, especially in the axils of the ribs, with rufous pubescence. The small white flowers are produced on long slender pedicels in many-flowered umbels arranged in terminal compound flat-topped panicles with long radiating branches, which are sometimes two feet in diameter; they appear in August and September, and are very conspicuous as they rise above the dark green foliage, giving to this fine tree an appearance entirely unlike that of any other inhabitant of northern forests.

Acanthopanax ricinifolium is common in Saghalin and Yezo, and I saw it occasionally on the mountains of central Hondo, where, however, it does not grow to the great size it attains in the forests of Yezo; here it is associated with Lindens, Magnolias, White Oaks, Birches, Maples, Cercidiphyllum, Walnuts, Carpinus, and Ostrya. The wood is rather hard, straight-grained, and light brown, with a fine satiny surface. In Yezo it is highly valued, and is used in considerable quantities in the interior finish of houses, and for furniture, cases, etc.

The illustration (Plate xvi.) is made from a photograph taken two years ago on the wooded hill near Sapporo, and represents a large although by no means an exceptionally large or remarkable specimen. At the right of the Acanthopanax two young Magnolias have sent up their trunks in search of light, and on the left appear a number of stems of the noble Japanese Grape-vine (Vitis Coignetiæ), which have climbed into its upper branches.

THE CORNELS, HONEYSUCKLES, AND PERSIMMONS, THE STYRAX FAMILY, THE ARBORESCENT MEMBERS OF THE HEATH FAMILY, THE ASHES, AND THEIR ALLIES.

Cornus, which is exceedingly common in North America, where sixteen or seventeen species are distinguished, is less abundant in Japan than in the other great natural botanical divisions of the northern hemisphere. In the northern regions of eastern America different species of Cornus often form a considerable part of the shrubby undergrowth which borders the margins of the forest or lines the banks of streams, lakes, and swamps. In Japan these shrubby species, or their prototypes, do not exist. High up among the Nikkō Mountains, on rocks under the dense shade of Hemlocks, we saw a few dwarf sprawling plants of the Siberian and north China Cornus alba, but did not encounter in any other part of the empire a shrubby Cornel. High up on these mountains, too, the ground is carpeted with the little Bunch-berry, the Cornus Canadensis of our own northern woods, which is also common in some parts of Yezo, and on the Kurile Islands, where a second herbaceous Cornel, with large white floral scales, Cornus Suecica, is found. This is a common plant, too, in all the boreal regions of North America from Newfoundland and Labrador to Alaska, and in northern Europe and continental Asia. Of arborescent Cornels the flora of Japan possesses only two species, Cornus Kousa and Cornus macrophylla, and neither of these is endemic to the empire.

Cornus Kousa represents in Japan the Cornus florida of eastern America and the Cornus Nuttallii of the Pacific states. From these trees it differs, however, in one particular; in our American Flowering Dogwoods, the fruits, which are gathered into close heads, are individually distinct, while in the Japan tree and in an Indian species they are united into a fleshy strawberry-shaped mass, technically called a syncarp. Owing to this peculiarity of the fruit, botanists at one time considered these Asiatic trees generically distinct from the American Flowering Dogwoods, and placed them in the genus Benthamia, which has since been united with Cornus. In Japan, Cornus Kousa is apparently not common; certainly it is not such a feature of the vegetation in any part of the empire which we visited as Cornus florida is in our middle and southern states. Indeed, we only saw it in one place among the Hakone Mountains, and on the road between Nikkō and Lake Chuzenji, where it was a bushy, flat-topped tree, not more than eighteen or twenty feet high, with wide-spreading branches. The leaves are smaller and narrower than those of our eastern American Flowering Dogwood; the involucral scales are acute and creamy white, and the heads of flowers are borne on longer and much more slender peduncles. Cornus Kousa also inhabits central China; it was introduced into our gardens several years ago, and it now flowers every year in the neighborhood of New York, where it was first cultivated in the Parsons' Nursery at Flushing. As an ornamental plant it is certainly inferior in every way to our native Flowering Dogwood, and in this country at least it will probably never be much grown except as a botanical curiosity.

The second arborescent Japanese Cornel, Cornus macrophylla, often known by its synonym, Cornus brachypoda, is also an inhabitant of the Himalayan forests, where it is common between 4,000 and 8,000 feet above the sea-level, and of China and Corea. It is one of the most beautiful of the Cornels, and in size and habit the stateliest and most imposing member of the genus. In Japan, trees fifty or sixty feet in height, with stout well-developed trunks more than a foot in diameter, are not uncommon, and when such specimens rise above the thick undergrowth of shrubs which in the mountain regions of central Japan often cover the steep slopes which descend to the streams, they are splendid objects, with their long branches standing at right angles with the stems, and forming distinct flat tiers of foliage, for the leaves, like those of our American Cornus alternifolia, are crowded at the ends of short lateral branchlets which grow nearly upright on the older branches, so that in looking down on one of these trees only the upper surface of the leaves is seen. These are five to eight inches long and three or four inches wide, dark green on the upper surface, but very pale, and sometimes nearly white, on the lower surface. The flowers and fruit resemble those of Cornus alternifolia, although they are produced in wider and more open-branched clusters; and, like those of this American species, they are borne on the ends of the lateral branchlets, and, rising above the foliage, stud the upper side of the broad whorls of green.

Cornus macrophylla is exceedingly common in all the mountain regions of Hondo, where it sometimes ascends to 4,000 feet above the sea, and in Yezo, where it is scattered through forests of deciduous trees, usually selecting situations where its roots can obtain an abundant supply of moisture. This fine tree was introduced into the United States many years ago through the Parsons' Nursery, but I believe has never flourished here. In the Arnold Arboretum, where numerous attempts to cultivate it have been made, it has never lived more than a few years at a time. Raised from seed produced in the severe climate of Yezo, Cornus macrophylla may, however, succeed in New England, where, if it grows as it does in Japan, it will prove a good tree to associate with our native plants.

Cornus officinalis, as it was first described from plants found in Japanese gardens, has usually been considered a native of that country. But, although it has been cultivated in Japan for many centuries on account of its supposed medical virtues, it is probably Corean. It may be considered, perhaps, a mere variety of the European and Asiatic Cornelian Cherry, Cornus Mas, from which the Corean tree is best distinguished by the tufts of rusty brown hairs which occupy the axils of the veins on the lower surface of the leaves. In the Botanic Garden in Tōkyō, which includes the site of a physic-garden established in the early days of the Tokugawa dynasty, there is a group of trees of Cornus officinalis, which appear to have attained a great age; they are bushy plants, perhaps thirty feet tall, with bent and twisted half-decayed trunks, and contorted branches, which form broad thick round heads, and in October were loaded with the bright Cherry-like fruit.

The Honeysuckle family is represented in Japan by seven genera and a large number of species, especially of Viburnum; but none of them can be considered trees, although Diervilla Japonica is occasionally almost arborescent in size and habit. The Japanese Viburnums are now all pretty well known in our gardens, with the exception of Viburnum furcatum, a common northern and mountain plant, so similar to our American Hobble-bush, Viburnum lantanoides, that some authors have considered the two plants identical, and Viburnum Wrightii, a distinct, black-fruited species of northern Japan, where the American botanist, Charles Wright, detected it when the Wilkes Expedition explored the shores of Volcano Bay. Viburnum furcatum is distributed through the mountain regions of the empire, and is one of the commonest species. Sometimes it grows to the height of fifteen feet; and it is always conspicuous from its great thick reticulate-veined, nearly circular leaves, which, in the autumn, turn to marvelous shades of scarlet, or to deep wine-color. If this fine plant takes kindly to cultivation it will prove a real acquisition to our gardens.

Ericaceæ abound in Japan, where we miss, however, such familiar American types as Kalmia, Oxydendrum, and Gaylussacia. Vaccinium is multiplied in species; but, with the exception of the red-fruited Vaccinium Japonicum and the black-fruited Vaccinium ciliatum, they are not very abundant, and are mostly confined to alpine summits, where the species are found which, in the extreme north, encircle the globe; and Blueberries nowhere cover the forest-floor with the dense undergrowth which is common in our northern woods. The broad-leaved evergreen true Rhododendrons are not very common in Japan, where there are only two species, and, being mostly confined to high elevations, they nowhere make the conspicuous feature in the landscape which Rhododendron maximum produces in the valleys of the southern Alleghany Mountains, or Rhododendron Catawbiense makes around the summit of Roan Mountain, in North Carolina and Tennessee. Most of the Japanese Azaleas produce purplish or brick-colored flowers; and in spite of all that travelers have said of the splendor of Japanese hillsides at the time when the Azaleas are in bloom, it is doubtful if they compare in beauty with some Alleghany mountain-slopes when these are lighted up with the flame-colored flowers of Azalea calendulacea, or with the summit of Roan Mountain during the last days of June, when one of the greatest flower shows of the world is spread there for the admiration of travelers.

None of the Japanese Azaleas, excepting, perhaps, Rhododendron Sinense, the Azalea mollis of gardens, produce such beautiful flowers as those of American species like Rhododendron (Azalea) viscosum, Rhododendron nudicaule, or Rhododendron arborescens. None of the Japanese Rhododendrons can be considered trees, although one or two of the deciduous-leaved species grow to the height of twenty or, possibly, thirty feet.

Andromeda Japonica, now common in our gardens, is properly a tree, for in the temple park of Nara, where it grows in profusion, there are specimens at least thirty feet in height, with stout well-formed trunks six or eight feet in length. Enkianthus campanulatus, the representative of a small genus of southern and eastern Asiatic trees, may be expected to become an ornament in our gardens of much interest and beauty; and as it grows as far north as the shores of Volcano Bay in Yezo, and up to over 5,000 feet in central Hondo, it may flourish in the climate of New England. Enkianthus campanulata is a slender bushy tree, sometimes thirty feet in height, with a smooth light red trunk, occasionally a foot in diameter, and thick smooth round branchlets. The leaves are mostly oval, sharply serrate, firm, dark green above and pale yellow-green below, about three inches long, and one inch wide; they are deciduous, and in the autumn, before falling, turn clear light yellow. The flowers are campanulate, pure white, and are borne on slender stalks in many-flowered drooping racemose panicles. By Japanese botanists it is spoken of as one of the most beautiful flowering trees in Japan, and we considered ourselves fortunate in securing a supply of ripe seed, for this species is still very rare in cultivation. There is but one other Japanese plant of the Heath family which can pass as a tree; this is the handsome Clethra canescens, or, as it is more generally known in Japan, at least, Clethra barbinervis, a more recent name. It is a beautiful small tree, occasionally twenty-five or thirty feet in height, with a slender trunk, a narrow oblong head, long-stalked obovate pointed leaves, four to six inches in length, and very dark green on the upper surface, and pale on the lower surface with hoary pubescence, which also covers the branches of the inflorescence and the outer surface of the calyx of the flowers. These are white, and are produced in slender upright terminal panicled racemes six to twelve inches long, and open in succession during several weeks in August and September. In southern Yezo Clethra canescens grows nearly down to the sea-level, and on the mountains of the southern islands; in central Hondo, where it is a common forest-plant, growing usually near the borders of streams and lakes, it reaches an elevation of over 5,000 feet, so that there is reason to believe that this fine species will thrive in our climate if plants are raised from seed produced at high elevations, although up to the present time those which have been sent to the Arnold Arboretum have never been very satisfactory. Clethra canescens grows, not only in Japan, but in China, Java, the Philippines, and Celebes.

Although we have learned to look upon Japan as the home of the Persimmon, which is intimately associated with the expression of modern Japanese art, it is doubtful if either of the species of Diospyros commonly encountered in that country is really indigenous in the empire, where they were both probably introduced, with many other cultivated plants, from China. The more common and important of the two species is, of course, the Kaki, Diospyros Kaki, which is planted everywhere in the neighborhood of houses, which in the interior of the main island are often embowered in small groves of this handsome tree. In shape it resembles a well-grown Apple-tree, with a straight trunk, spreading branches which droop toward the extremities and form a compact round head. Trees thirty or forty feet high are often seen; and in the autumn, when they are covered with fruit, and the leaves have turned to the color of old Spanish red leather, they are exceedingly handsome. Perhaps there is no tree, except the Orange, which, as a fruit-tree, is as beautiful as the Kaki. In central and northern Japan the variety which produces large orange-colored ovate thick-skinned fruit is the only one planted, and the cultivation of the red-fruited varieties with which we have become acquainted in this country is confined to the south. A hundred varieties of Kaki, at least, are now recognized and named by Japanese gardeners, but few of them are important commercially in any part of the country which we visited, and, except in Kyōto, where red kakis appeared, the only form I saw exposed for sale was the orange-colored variety, which, fresh and dried, is consumed in immense quantities by the Japanese, who eat it, as they do all their fruits, before it is ripe, and while it has the texture and consistency of a paving-stone.

Diospyros Kaki, or an allied species, is hardy in Peking, with a climate similar to that of

New England, and fully as trying to plant life; it fruits in southern Yezo, and decorates every garden in the elevated provinces of central Japan, where the winter climate is intensely cold. There appears, therefore, to be no reason why it should not flourish in New England if plants of a northern race can be obtained; and, so far as climate is concerned, the tree, which, in the central mountain districts of Hondo, covers itself with fruit year after year, will certainly succeed in all our Alleghany region from Pennsylvania southward. In this country we have considered the Kaki a tender plant, unable to survive outside the region where the Orange flourishes. This is true of the southern varieties which have been brought to this country, and which may have originated in southeastern Asia, in a milder climate than that of southern Japan, for the Kaki is a plant of wide distribution, either natural or through cultivation. But the northern Kaki, the tree of Peking and the gardens of central Japan, has probably not yet been tried in this country. If it succeeds in the northern and middle states, it will give us a handsome new fruit of good flavor, easily and cheaply raised, of first-rate shipping quality when fresh, and valuable when dried, and an ornamental tree of extraordinary interest and beauty.

Diospyros Lotus, which is probably a north China species, and which is naturalized or indigenous in northwestern India, and naturalized in the countries bordering the Mediterranean, is occasionally cultivated in northern Japan, where, however, as it does not appear to be more hardy than the Kaki, it does not seem to be much esteemed. The fruit is small and of an inferior quality. Diospyros Lotus may be expected to endure the climate of our northern states.

In Japan, Styraceæ is represented by Symplocos with half a dozen species, all shrubs rather than trees, by Pterostyrax, which replaces our Mohrodendron, from which the Japanese genus only differs in its terminal paniculate inflorescence, five-parted flowers, and small fruits; and by Styrax with two species. Neither of the two species of Pterostyrax equals in size our Mohrodendron Carolinum, which, under favorable conditions, becomes a tree eighty to a hundred feet high on the southern Alleghany Mountains, and neither of them approaches our two arborescent Mohrodendrons in the beauty of their flowers, which, although produced in ample clusters, are individually small. Pterostyrax corymbosum, which I believe to be almost exclusively a southern species, I saw only in the Botanic Garden in Tōkyō, where there is a bushy plant eighteen or twenty feet in height. Pterostyrax hispidum, which is now beginning to be known in our gardens, where it is hardy from Boston to Philadelphia, is a bushy tree or shrub, which we saw wild in Japan only on the banks of a stream among the mountains above Fukushima, on the Nakasendō, where we found a single plant twenty or twenty-five feet in height.

As an ornamental plant, the most valuable of this family, as represented in Japan, is certainly Styrax Obassia, a tree which grows as far north as Sapporo, in Yezo, and which may therefore be expected to be as hardy as Cercidiphyllum, Syringa Japonica, Magnolia Kobus, or any of the Yezo trees, with which it grows and which flourish here in New England. Styrax Obassia, as it appears in Yezo and on the mountains of central Hondo, where it is common between 3,500 and 4,000 feet above the sea, is a tree twenty to thirty feet in height, with a straight slender stem, long and graceful branches well clothed with nearly circular leaves dark green on the upper surface, pale on the lower surface, and often more than six inches across. The white bell-shaped beautiful flowers, which are nearly an inch in length, are borne in long drooping racemes, and are produced in the greatest profusion. The second species, Styrax Japonica, is a common plant in the mountain forests of Hondo and in southern Yezo, and is a shrub or occasionally a small tree twenty to thirty feet high. It is now well known in American and European gardens.

From the Olive family we miss, in Japan, Forestiera, an exclusively American genus, and Chionanthus, which is eastern American and Chinese. Fraxinus and Osmanthus are common to the floras of Japan and eastern America, and in Japan, Ligustrum and Syringa, both Old World genera, are represented. In eastern America, Fraxinus appears in nine species; in Japan there are probably not more than two indigenous species, and only Fraxinus longicuspis is endemic. This is a tree of the Ornus section, which is rather common in the elevated Hemlock forests of Hondo, and ranges northward into Yezo. It is a slender tree, twenty to thirty feet in height, with thin rigid ashy gray branchlets, black buds, and leaves with five stalked ovate acute and finely or coarsely serrate leaflets, which in the autumn are conspicuous from the deep purple color to which they change.

Fraxinus Manchurica, which is also common in Manchuria, Saghalin, and Corea, is a noble tree in Yezo, where it is exceedingly abundant in low ground near the borders of swamps and streams, and where it often rises to the height of a hundred feet, and forms tall straight stems three or four feet in diameter; its stout orange-colored branchlets, large black buds, ample leaves, with lanceolate acute coarsely serrate leaflets, and great clusters of broad-winged fruits, well distinguish this species, which is certainly one of the noblest of all the Ashes, and one of the most valuable timber-trees of eastern Asia. For many years Fraxinus Manchurica has inhabited the Arnold Arboretum, where it is hardy, and where it promises to grow to a good size. Another Ash-tree commonly cultivated along the borders of Rice-fields near Tōkyō is referred by the Japanese botanists to the Fraxinus pubinervis of Blume. This has every appearance of being an introduced tree in Japan; but I was unable to obtain fruit or any satisfactory information with regard to it.

Syringa Japonica is rather common in the deciduous forests on the hills of central Yezo, and I saw it occasionally on the high mountains of Hondo. In its native country, the Japanese Lilac, when fully grown, is an unshapely straggling tree, twenty-five to thirty feet in height, with a trunk rarely twelve or eighteen inches in diameter, and does not display the beauty of foliage and the compact handsome habit which we associate with this plant in our New England gardens, where it is far more beautiful than in its native forests.

Osmanthus Aquifolium, or, as it is more commonly called in gardens, Osmanthus ilicifolius, is usually supposed to be a Japanese tree. I saw it in city gardens, and in greater perfection in the mountain region of central Hondo, where it is often planted near dwellings and by the roadside, and where it sometimes grows to the height of thirty feet, and makes a trunk a foot or more in diameter, and a broad compact round head, loaded in October with fragrant flowers. But where I saw it, it had evidently been planted, and if it is a Japanese species, which is doubtful, it is only indigenous in the extreme south.

Ligustrum is poorly represented in Japan, and of the three species found within the borders of the empire only the evergreen Ligustrum Japonicum of the south becomes a tree. Of the other species, Ligustrum medium is much more common than Ligustrum Ibota; at the north it is found in moist low forests, but farther south ascends to high elevations, where, in central Japan, Ligustrum Ibota is also found.

In the remaining Gamopetalous orders of trees Japan possesses only Ehretia acuminata, which inhabits the Loochoo Islands, and possibly reaches the southern shores of Kûyshû, a small tree, which I saw only in the Botanic Garden of Tōkyō, and the beautiful Clerodendron trichotomum, which in late summer enlivens the banks of streams with its great masses of tropical foliage and brilliant flowers, and in Yezo often attains to the size and habit of a small tree.

THE LAUREL, EUPHORBIA, AND NETTLE FAMILIES.

The American traveler landing for the first time in Yokohama is surprised at the abundance of arborescent Lauraceæ, which here, with evergreen Oaks and Celtis australis, make the principal features of the woods which cover the coast-bluffs and surround the temples. The most abundant of the Lauraceæ in this part of Japan appears to be the Camphor-tree, Cinnamomum Camphora, one of the most beautiful of all evergreens, and probably indigenous in southern Japan. In that part of the country which we visited, however, it had every appearance of having been planted. Even at Atami, on the coast some distance below Yokohama, a popular winter resort famed for the mildness of the climate and for the geyser, which attract many visitors, the Camphor-tree is probably not indigenous. Near the town is the grove of Kinomiya, where may be seen what is popularly supposed to be the largest Camphor-tree in Japan. It is really a double tree, as the original stem has split open, leaving irregular faces, which have become covered with bark. Between the two parts there is sufficient space for a small temple. The larger of the two divisions at five feet from the surface of the ground, and well above the greatly swollen base, girths thirty-three feet eight inches, and the smaller twenty-seven feet six and a half inches. This remarkable tree, which has every appearance of great age, is still vigorous and in good health. Atami is celebrated for the skill of its workers in wood, and for the production of many small articles made from the wood of the Camphor-tree. The hills along the coast are covered with groves of Orange-trees, and in the temple gardens were many southern trees which we did not see in perfection in other parts of the empire.

On the coast of this part of Japan, Cinnamomum pedunculatum becomes a tree thirty or forty feet in height, and on the neighboring Hakone Mountains ascends to elevations of a couple of thousand feet. It is a handsome tree, with ample ovate acute lustrous leaves, pale or nearly white on the lower surface, and long-stalked flowers and fruit. In the same region two other arborescent Lauraceæ grow naturally — Litsea glauca and Machilus Thunbergii. They are both evergreens, and are handsome trees, especially the Litsea, which bears oval leaves pointed at both ends, silvery white on the lower surface, and often six inches long, and near the ends of the branches abundant clusters of black fruit. These two trees are the most northern in their range of the Lauraceæ of Japan with persistent foliage, and they may be expected to thrive in this country where the evergreen Magnolia and the Live Oak flourish.

The flora of eastern Asia is rich in Linderas, no less than twenty species having already been found in the Chinese empire and in Corea, while in North America there are only two — Lindera Benzoin, the common Spice-bush of northern swamps, and the southern Lindera melissæfolia. Japan possesses half a dozen indigenous Linderas, although none of them are endemic, and in the mountain regions of Hondo several species are common, and make notable features in the shrubby growth which covers hillsides and borders streams and lakes.

PLATE XVII.

C. F. Faxon del.

LINDERA TRILOBA, Blume.

The most beautiful, perhaps, of the Japanese species is Lindera umbellata, a southern plant, found also in central China, which I saw only in the Botanic Garden at Tōkyō, where it forms a stout bush eight or ten feet high. The leaves, which appear in the spring with the flowers, are lanceolate-acute, very gradually narrowed at the base, rounded at the apex, entire, and often six or eight inches in length; they are lustrous on the upper surface, and pale and covered on the midribs and veins on the lower surface with rufous pubescence. The fruit, which is a quarter of an inch in diameter and brilliant scarlet, is produced in great quantities in dense axillary clusters on the branches of the previous year, and ripens in August and September. As I remember it, this seems one of the most beautiful plants which I saw in Japan. It may be expected to thrive in the southern states and in southern Europe, but it will probably not be able to support the cold of the north.

As a garden-plant for this region, Lindera obtusiloba is, perhaps, the most promising; and we were fortunate in securing a sufficiently large quantity of seeds, gathered at high elevations in central Hondo, to give it a good trial. Lindera obtusiloba (see Plate xvii.) often becomes a bushy tree, twenty to twenty-five feet in height, with a short stout trunk, terete brown branchlets, and conspicuous winter-buds covered with imbricated chestnut-brown scales. The leaves appear with the flowers and are broadly ovate, palmately three-nerved, mostly three-lobed at the apex, three or four inches long and broad, thick and firm, lustrous above, and pale and often puberulous on the veins below. In the autumn, before falling, they turn to a beautiful clear yellow color, and make a handsome contrast with the shining black fruits, which are borne on hairy stalks in few-fruited axillary clusters, produced on short spur-like lateral branchlets of the previous year. This handsome plant, although it grows to its largest size in central Hondo at four or five thousand feet above the sea, does not, so far as we observed, range north of the Nikkō Mountains, and, therefore, does not reach Yezo, where only Lindera sericea is found. This is a small slender shrubby species, with precocious flowers, oval entire pointed leaves, silky-canescent at first, and at maturity dark green on the upper and pale on the lower surface, and small black fruit.

The other species of Lindera, which may possibly prove hardy in our northern gardens, are Lindera triloba and Lindera præcox. The first is a common plant in Hondo, where it does not, however, ascend to the heights reached by Lindera obtusiloba, which is a more northern and a hardier plant. Lindera triloba often grows twenty feet tall, and produces trunks six inches in diameter, from which spring numerous slender divergent branches well clothed with leaves. These appear with the flowers, and are elliptical or oblong, wedge-shaped at the base, and divided at the apex into three acute lobes separated by deep broad sinuses rounded in the bottom; they are three-nerved, membranaceous, light green above, pale, and covered below on the ribs with rufous pubescence, three or four inches long, and two or three inches broad, and are borne on slender petioles. The fruit is half an inch in diameter, and is produced in few-fruited umbels on short stout club-shaped stalks.

Lindera præcox, like our American species, flowers before the leaves appear; it is a bushy tree fifteen to twenty feet in height, with stout divergent light brown branches, and is conspicuous in midsummer from the large size of the flower-buds, which are then fully grown, and which probably open during the winter or in earliest spring. The leaves are ovate, long-pointed, rather thin, dark green above, pale and often pubescent below, and two or three inches in length, with long slender stalks. The fruit, which is nearly an inch in diameter, is reddish brown and marked with many small white dots; the flesh is thin, papery, and very brittle. Lindera præcox is common on the Hakone Mountains; we found it near Agematsu, on the Nakasendō, and Mr. Veitch collected it on Mount Chōkai-zan, on the northwest coast of Hondo. If it inhabits the Nikkō Mountains we missed it there, and also on Hakkoda, near Aomori, where Lindera sericea was the only species seen. The other Japanese species, Lindera glauca, is a southern black-fruited plant with precocious flowers and the habit and general appearance of Lindera sericea, from which it differs in its larger leaves and more rigid branches.

Of Elæagnus, the sole representative of its family in Japan, we only saw growing naturally Elæagnus umbellata, a variable plant in the size and shape of its leaves and fruit, and one of the commonest shrubs in Japan from the level of the sea up to elevations of 5,000 feet. In the mountainous regions and at the north it is often planted near houses for the sake of its small acid fruit; and in cultivation it not infrequently rises to the size and dignity of a small tree. Elæagnus umbellata is now well established in our gardens, where it flowers and fruits as freely as it does in Japan.

The now well-known Elæagnus longipes was often seen in gardens, especially among the mountains and in Yezo, but we did not notice it growing wild. In old age it sometimes attains the height of twenty or twenty-five feet, and forms a stout straight trunk a foot in diameter. Such a plant, evidently of great age, may be seen in the Botanic Garden at Tōkyō. The beautiful Elæagnus pungens, with its long wand-like stems, now a familiar object in several varieties in the gardens of southern Europe, was seen in the temple grounds at Nara and by the roadside near Kyōto, where it appeared to have escaped from cultivation rather than to be an indigenous plant. Of the other reputed Japanese species we could hear nothing

The flora of Japan, although it is comparatively rich in Euphorbiaceæ, does not contain any important trees belonging to this family. One species of Daphniphyllum, a Malayan genus with beautiful lustrous evergreen foliage and handsome fruit, and now known in the gardens of temperate Europe in the shrubby Daphniphyllum glaucesens, attains the size of a small tree; this is Daphniphyllum macropodum, which we saw not far from Gifu, growing, as it seemed, naturally. It is interesting to note that one species of this tropical genus, Daphniphyllum humile, grows far north in Yezo, where, as well as on the mountains of northern Hondo, it is a common under-shrub in the forest of deciduous trees. We obtained a supply of seeds of this handsome plant, although it is hardly to be expected that it will be able to survive our northern winters, as it will miss here the continuous covering of snow under which it is buried in Yezo during many months of the year.

Of the small genus of Aleurites of eastern tropical Asia and the Pacific islands, one species reaches southern Japan, Aleurites cordata, which we saw only in the Botanic Garden at Tōkyō. This little tree has large long-stalked three-lobed leaves, inconspicuous flowers in terminal panicles, and large black drupe-like fruit; it may be expected to grow in the southern

PLATE XVIII.

ULMUS CAMPESTRIS, L.

states; but it will be valued for its botanical interest, and not for its beauty. And this is true of the other Japanese tree of the Euphorbia family, Excœcaria Japonica, which may possibly prove hardy here in New England, as we found it growing on high elevations on the Nakasendō near Agematsu in central Japan, as well as on the high Otome-toge in the Hakone Mountains; and Mr. Veitch gathered specimens on Mount Chōkai-zan, on the northwest coast of Hondo. It is a small tree with thick firm dark green leaves which vary from oval or obovate to obovate-lanceolate, and are sometimes six or seven inches long, and three-lobed fruit three quarters of an inch in diameter.

Of the Nettle or Elm family, Japan possesses some important trees, although in Elms themselves the flora of Japan is poor as compared with that of eastern North America, where there are five well-distinguished species, while in Japan there are only two; these are both continental, reaching in Japan their most eastern home. In Hondo Elm-trees are not common, and in that island are nowhere such features of vegetation as they are in our New England and middle states and in Europe, and it is only in mountain forests between 3,000 and 5,000 feet above the sea-level that occasional small plants of Ulmus campestris, with branchlets often conspicuously winged, appear. In Yezo, however, this tree is much more abundant, growing on the river-plains nearly at the sea-level and in the forests which cover the low hills, not infrequently becoming a prominent feature of the landscape. In Sapporo, where many fine old specimens were left in the streets by the American engineers who laid out the town, individuals seventy or eighty feet tall, with trunks three or four feet in diameter, may be seen. The broad heads of graceful pendent branches reminded us of New England, for this Japanese form of the Old World Elm has much of the habit of the American White Elm. The portrait of one of these trees (see Plate xviii.), although not a large one, growing a mile or two from Sapporo, gives a fair idea of the habit of this tree in Yezo.

The second Elm of Japan grows in all the mountain-woods near Sapporo. The Russian botanists have considered it a variety of Ulmus scabra, to which the name laciniata has been given, and which is principally distinguished by the peculiar shape of the leaves, which are coarsely serrate, often six or seven inches long, three or four inches broad, and three-lobed at the wide apex. It is a small tree, barely more than thirty feet tall, as we saw it, and very fragrant, like our American Slippery Elm, which in habit it much resembles. It is from the tough inner bark of this tree that the Ainos weave the coarse brown cloth from which their clothes are made. The process is a simple one; the bark is stripped from the trees in early spring, and is then soaked in water until the bast, or inner bark, separates from the outer in long strips, which are twisted by the women into threads, and are then ready for use. This interesting tree is not in cultivation, I believe, and we reached Yezo too late to obtain its seeds. It is desirable, however, that it should be brought into our gardens, not only on account of the curious appearance of the leaves, but that its development may be watched; for when it can be compared in a living condition with the European and Siberian forms of Ulmus scabra, it may prove sufficiently distinct to be regarded as a species.

The Keaki, Zelkova Keaki, a member of this family, is, perhaps, the largest deciduous-leaved tree of Japan; it is its most valuable timber-tree. The Keaki may be described as a Beech, with the foliage of an Elm. The bark is smooth and pale, like the bark of a Beech-tree, and the dense compact round head of slender branchlets resembles the crown of that tree, while the leaves, which are large, ovate-acute, coarsely serrate, and roughened on the upper surface, are like those of the Elm.

Zelkova is a genus with three arborescent species; one inhabits Crete, another the Caucasus, and the third Japan. The flowers are very similar to those of the Elm, and are unisexual or rarely polygamous, and are produced in early spring on branchlets of the previous year, the males clustered in the axils of the lower leaves, and the females solitary in those of the upper leaves. The fruit is a small drupe, more or less irregularly oblique in shape and two-beaked with the remnants of the eccentric style; it has a membraneous or slightly fleshy outer covering and a thin hard endocarp or stone, containing a single compressed concave horizontal seed, without albumen, the thick embryo filling its cavity.

Zelkova Keaki sometimes grows to the height of a hundred feet, and produces a trunk eight to ten feet in diameter. Such specimens are often found in the gardens surrounding temples in the large cities, and by village roadsides in the interior provinces. If any wild Keakis are left in the forests of Japan they must be rare, and I am not sure that we saw this tree growing naturally, although it is everywhere one of the most commonly planted deciduous trees. Large specimens, which we saw on the Nakasendō, near Agematsu, in one of the mountain provinces of central Hondo, and a very remote region, may have been growing naturally; but even this is doubtful, for the Nakasendō has been a traveled highway for at least twelve hundred years, and a thousand years ago was probably more frequented than it is now. Of the range and habitat of this tree I have, therefore, no idea whatever.

The wood is more esteemed by the Japanese than that of any of their other trees. It is noted for its toughness, elasticity, and durability, both in the ground and when exposed to the air; it is considered the best building material in Japan, although it has become so scarce and expensive that keaki is not now used for this purpose, except in temples, where the large round light brown highly polished columns which support the roof are always made of this wood. It is still much used in cabinet-making and turnery, and in the manufacture of many small articles, which always command high prices.

Zelkova Keaki is probably the only Japanese tree which is worth introducing into this country on a large scale as a timber-tree; that it will thrive here at least as far north as southern New England, the plants in Dr. Hall's garden in Warren, Rhode Island, indicate. There are several of those raised from seed sent home by Dr. Hall in 1862; they have received no special care, the soil in which they were planted is not exceptionally good, and their growth has been no doubt checked by overcrowding. They are now, however, at least fifty feet high, and have produced trunks a foot in diameter; they flowered and fruited last year, and our illustration (see Plate xix.) is made from specimens taken from these trees, with the exception of the large single leaf, which has been drawn from a specimen gathered in Japan.

The Zelkova, of all Japanese trees, should be better known in eastern America, where it may, perhaps, become an imported timber-tree, and produce wood as strong as our best oak, which it surpasses in compactness, durability, and lightness, for keaki, in comparison with its strength, is remarkably light.

ZELKOVA KEAKI, Sieb.

There is little to be said of the other Japanese trees of the Elm family. Celtis Sinensis, with its thick coriaceous leaves and dull red berries, is one of the first trees to greet the traveler landing in Yokohama, where it is common in the groves which cover the shore-bluffs, growing with the Camphor-tree and the evergreen Oaks. It is a southern species of wide range in southeastern Asia, which we cannot hope to grow in this country, except in the southern states.

Aphananthe aspera, a Celtis-like tree with ample bright green leaves and black fruit, ranges as far north as central Yezo, and may be expected to give interest and variety to dendrological collections in the United States and Europe, although to the mere lover of trees with peculiar foliage or with showy flowers and fruit it will not appear sufficiently distinct from our native Nettle-tree.

Of the Broussonetias or Paper Mulberries, of which two or three species are included in the flora of Japan, I saw specimens only in the Botanic Garden at Tōkyō. They are all trees of the south, or more probably introductions from China. The White Mulberry, Morus alba, however, is certainly a Japanese species, as it grows as a small tree in the remote and primeval forests of Yezo, although the numerous forms cultivated by the Japanese as food for the silkworm are usually of Chinese origin. Of the Fig-trees which appear in the flora of Japan, I saw nothing at all, with the exception of one or two cultivated shrubby plants. They all belong to the extreme south, and inhabit regions we did not visit.

THE WALNUTS, BIRCHES, ALDERS, AND HORNBEAMS.

IN nut-bearing trees the forests of Japan are poor in comparison with those of eastern North America. The Hickory, if it ever existed in the ante-glacial forests of Asia, has entirely disappeared from them, and the Walnut family now appears in Japan in three genera — Juglans, Pterocarya, and Platycarya; the last two belong exclusively to the Old World. In Japan, Juglans is represented by Juglans Sieboldiana, a common forest-tree in Yezo and in the mountain regions of the other islands. As a timber-tree it is much less important than either of the two eastern American Walnuts, as specimens more than fifty feet high are uncommon; it is a wide-branched tree, resembling our Butternut in habit and in the color of its pale furrowed bark, as it does in the pubescent covering of the young branches, the lower surface of the leaves, and the fruit. The nuts are arranged in long racemes, and resemble those of the Asiatic, or, as it is familiarly called in commerce, the English Walnut (Juglans regia), rather than our American walnuts, which are deeply sculptured into narrow ridges, while the surface of the Japanese nut is smooth, or sometimes more or less pitted; it is pointed at the apex with thickened wing-like sutures, and is often an inch and a half long and about an inch broad, although it varies considerably both in size and shape; in flavor the kernel resembles that of the English walnut. The walnut is evidently an important article of food in Japan, as the nuts are exposed for sale in great quantities in the markets of all the northern towns. Juglans Sieboldiana is perfectly hardy here in New England, where it ripens its fruit; it is not worth growing, however, as an ornamental tree, as the Black Walnut surpasses it in size and beauty. It will produce fruit, however, in regions of greater winter cold than the English Walnut can support, and as a fruit-tree it may find a place in northern orchards, although the abundance and cheapness of English walnuts seem to forbid its cultivation as a source of profit.

I am unable to throw any light upon the curious Juglans cordiformis of Maximowicz, distinguished by its flattened, long-pointed, and more or less heart-shaped nuts. The tree which produces these peculiar nuts is not recognized by the Japanese botanists, who consider them extreme varieties of their common walnut. I looked in vain for nuts of this form in the markets of Hakodate, where they were first seen by the Russian naval officer Albrecht; afterward, however, I found them offered for sale by the Nurserymen's Association of Yokohama, and was told that they were collected on the sides of Fuji-san. A plant raised from one of these heart-shaped nuts has been growing for a number of years in the Arnold Arboretum, and has produced fruit. In habit and in foliage it is not distinguishable from plants of the same age of Juglans Sieboldiana. Juglans regia, although included in most works on the flora of Japan, is not a native of the empire; it is occasionally cultivated in the neighborhood of temples and as a fruit-tree, but we saw no evidence of its being anywhere indigenous, and it is probable that it was introduced from northern China, where one form of this tree apparently grows naturally.

THE WALNUT FAMILY.

Pterocarya, the curious genus with leaves like those of a Hickory, and long slender spikes of small hard nut-like fruits surrounded by foliaceous bracts, appears in Japan with one species; a second inhabits China, and a third, the type of the genus, the Caucasus. The Japanese Pterocarya rhoifolia is a large and important timber-tree. We first met with it on the lower margin of the Hemlock forest about Lake Yumoto, in the Nikkō Mountains, where it grows to no great size; and it was not until we ascended Mount Hakkoda, in the extreme northern part of Hondo, that we saw this fine tree to advantage. On the slopes of this mountain it is exceedingly common at elevations of from 2,500 to 4,000 feet above the sea, and, next to the Beech, is the largest deciduous tree of the region, often rising to a height of eighty feet, and producing trunks two and a half feet in diameter. It is a broad-topped tree, with stout branches, which spread nearly at right angles to the stem, and form a dense leafy crown. In winter the Japanese Pterocarya may be readily recognized by its orange-colored branchlets, thickly beset with small light-colored lenticels, and by the stout acute buds, three quarters of an inch long, covered with apiculate black puberulous scales conspicuously marked with clusters of pale hairs. The leaves are unequally pinnate, eight or ten inches long, and four to six inches broad, with stout hairy petioles, and six or seven pairs of lateral leaflets, which are acute, unequally rounded at the base, long-pointed, finely serrate, yellowish green, and covered on the lower surface of the midribs with pale or rusty brown pubescence. In the first days of October, when the fruit was fully ripe and just ready to drop, the leaves were beginning to turn yellow; a month later, in the forest above Lake Yumoto, the trees were bare of foliage. A specimen of the wood of Pterocarya rhoifolia, for which I am indebted to the officers of the Forestry Department at Aomori, is white, soft, very light, and straight-grained, with bands of open ducts marking the layers of annual growth; it might be mistaken at the first glance for a piece of our American white pine.

The other Japanese member of the Walnut family, Platycarya strobilacea, we saw only in the Tōkyō Botanic Garden, where there is a tree fifteen or twenty feet high, which two years ago was covered with the curious cone-like heads of fruit which distinguish this genus. In the mountain regions of Kyūshū it is said to become a large and stately tree. Platycarya is occasionally cultivated in the botanic gardens of southern Europe, but I am not aware that it is growing in the United States.

Myrica Gale, in a distinct pubescent form, is as common in low marshy ground in Yezo as it is in the same latitude in North America, and a second species of Myrica, akin to our Bayberry, inhabits the sandy coast, although it does not range far north of the thirty-fifth parallel. This is the handsome evergreen Myrica rubra, a small shapely tree, now well known in California gardens, and occasionally cultivated in the southern Atlantic states.

In Japan, as in all other temperate northern lands, the Cupuliferæ abound, and the deciduous forests of the northern islands are principally composed of Oaks, Beeches, Hornbeams, Alders, and Birches. The mountain forests of Hondo and those of Yezo contain many Birch-trees, which are also important elements of the forest in all northern and northeastern Asia. The Old World White Birch, Betula alba, in at least three of its forms, is common in central Yezo, and we saw also a number of trees of the typical form on the plains between Chuzenji and Yumoto, in the Nikkō Mountains. The most distinct of the Japanese forms of Betula alba is that which botanists call var. Tauschii, and which is distributed from southern Siberia through the Amour country, to Yezo, where it is a slender tree, sometimes eighty feet in height; it is distinguished by its larger and rather thicker leaves, which are of a deeper and more lustrous green on the upper surface than those of the other forms of the White Birch with which it is associated. It is certainly worth a place in our plantations. The variety verrucosa, well distinguished by the warts which beset the young branches, appears to be confined in Japan to Yezo, where, so far as we were able to observe, it is an exceedingly rare plant.

In the forests of Yezo, too, we saw, for the first time, Betula Maximowicziana. This is certainly one of the handsomest trees in Japan, and one of the most distinct and beautiful of the Birches; and its introduction into our plantations was alone well worth the journey to Japan. In Yezo, Betula Maximowicziana is a shapely tree, eighty or ninety feet in height, with a trunk two or three feet in diameter, covered with pale smooth orange-colored bark. Toward the base of old individuals the bark becomes thick and ashy gray, separating into long narrow scales. The branchlets are stout, covered with dark red-brown bark and marked by many pale lenticels. The leaves, however, are the most distinct feature of this tree; in size they are not equaled by those of any other Birch-tree, and as they flutter on their long slender stalks they offer a spectacle which can be compared with that which is afforded by our Silver-leaved Linden waving its branches before some Hemlock-covered hill of the southern Alleghany Mountains. The leaves of Betula Maximowicziana are broadly ovate, cordate at the base, coarsely and doubly serrate, very thin and membranaceous, dark green and lustrous on the upper surface, pale yellow-green on the lower surface, four to six inches long and four or four and a half inches broad. The flowers and fruit I have not seen. The male catkins in September are an inch and a half long, very slender, with bracts rounded and apiculate at the apex. From the seeds, for which I am indebted to the Forestry officials of Hokkaido, a large number of seedlings of this fine tree have been raised in the Arboretum. Specimens collected in the Nikkō Mountains by Mayr indicate that it is an inhabitant of Hondo, where, however, we did not see it. From Yezo it ranges northward through Saghalin into Manchuria. The tough thin bark is used by the Ainos for many domestic purposes.

The most common Birch of the high mountain forests of Hondo is Betula Ermani, a handsome species now well known in European and American collections, into which it has been introduced through the agency of the St. Petersburg Botanic Garden. In Hondo, where it is found scattered through the coniferous forests, it is common at elevations of from four to six thousand feet above the sea, and is conspicuous from the white bark of the trunk and the bright orange-colored bark of the principal branches. From the different forms of the White Birch this species can be readily distinguished in the herbarium by the long spatulate middle lobe of the bract of the female flower; in the forest the color of the bark of the branches well distinguishes it.

On the shores of Lake Yumoto we found a single individual of a black-barked Birch-tree, much like our American Betula lenta, with the same Cherry-like flavor in the bark of the branchlets. From Betula lenta it differed in its larger, more obtuse and paler winter-buds, in the more prominent midribs and veins of the leaves covered on their lower surface with silky

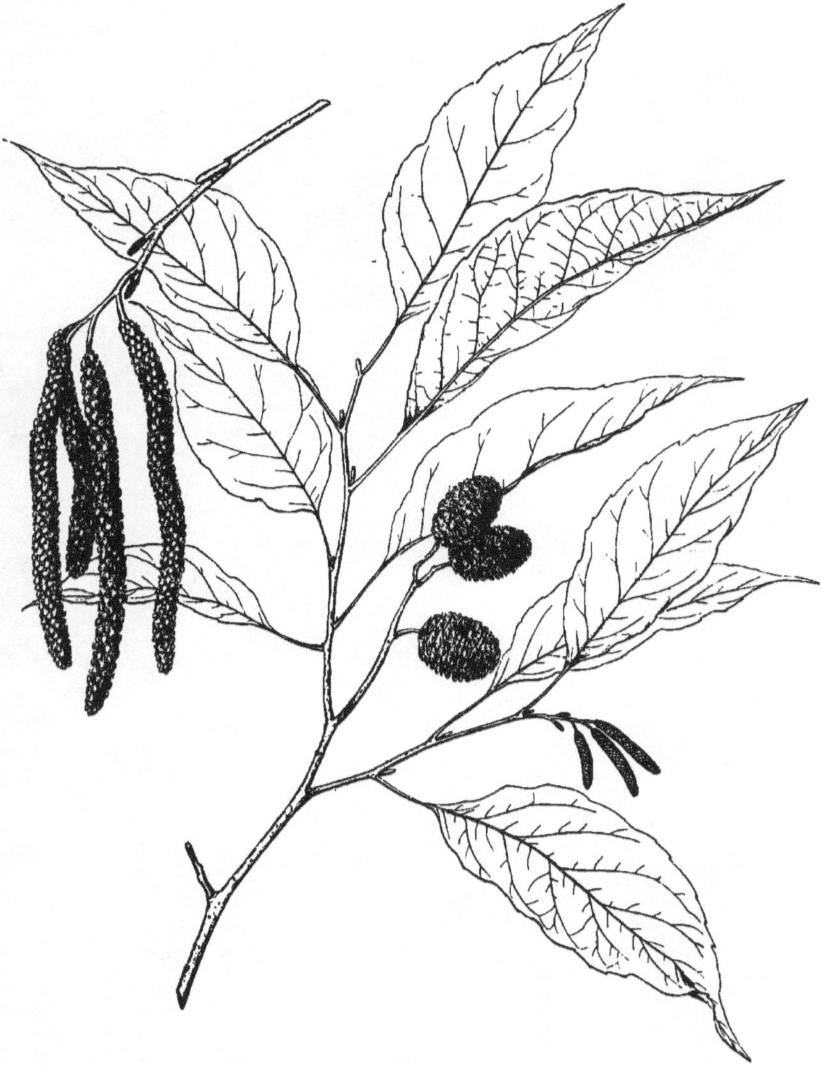

PLATE XX.

ALNUS JAPONICA, SIEB. ET ZUCC.

pubescence, and in the shorter cones of fruit, the lateral lobes of the bracts being narrow and acute, instead of broad and rounded, as in the American species. With considerable hesitation I have referred this tree to the Betula serra of Siebold & Zuccarini. The seedling plants which have been raised in the Arboretum will, perhaps, throw some light upon its true position. Betula ulmifolia, Betula Bhojpattra, and Betula corylifolia, included in the flora of Japan, we did not see.

In Japan Alders are more numerous in species, and grow to a much larger size than in eastern America. Alnus incana, which is only a shrub here, in Japan becomes in some of its forms a stately tree fifty or sixty feet in height, forming trunks often two feet in diameter. Trees of this size of the varieties glauca and hirsuta, the latter well characterized by the pale pubescence which covers the lower surface of the leaves, are common in Yezo, where they are found on low slopes in moist rich ground, but not often close to the banks of streams, which are usually occupied by Alnus Japonica. This is the largest and most beautiful of the Japanese Alders. It is a pyramidal tree, often sixty to eighty feet tall, well furnished to the ground with branches clothed with large dark green lustrous leaves. This species has been confounded with the rare North American Alnus maritima, from which it differs in habit and in the size and color of the leaves. The fruit of the two species is very similar, but the Japanese tree flowers in the spring, ripening its fruit in the autumn of the same year, while the American tree flowers in the autumn, and does not perfect its fruit until a year later. The figure of Alnus Japonica (see Plate xx.) has been made from a drawing of a wild specimen gathered in Yezo.

Alnus Japonica is sometimes found in our collections, and is generally cultivated under the name of Alnus firma. It is perfectly hardy in New England, where it grows rapidly, and promises to become a large and handsome tree. The true Alnus firma, which is largely planted along the margins of the Rice-fields near Tōkyō to afford support for the poles on which the freshly cut rice is hung to dry, was not seen growing under what appeared natural conditions; but the beautiful mountain-tree, distinguished by the thick conspicuously veined leaves, which has been considered a variety of Alnus firma (var. multinervis), we often saw on the mountains of Hondo, where it grows on dry rocky soil, and reaches elevations of some 5,000 feet above the sea-level. It is a graceful tree, sometimes twenty or thirty feet high, with slender spreading branches and thin flexible branchlets covered with ample thick dark green acute leaves with from sixteen to twenty-four pairs of pale conspicuous straight veins. When better known, this handsome tree will probably prove to be specifically distinct, and a garden-plant of value.

What has been considered a form of Alnus viridis (var. Sibirica) is a very distinct-looking plant in Japan, with broadly ovate cordate leaves fully twice as large as those produced by Alnus viridis in America or Europe. At high elevations on Mount Hakkoda we found it growing as a bushy tree from twenty to twenty-five feet tall, and forming a short stout trunk. A review of all the known forms of Alnus viridis will probably necessitate the separation of this Japanese plant from that species.

Of the Oak family it is in Carpinus only that the forests of eastern Asia are superior to those of America, where we have a single species of Hornbeam, a small tree confined to the eastern side of the continent. Europe possesses, also, a single species which extends to the Orient, where a second species is found. The forests which cover the Himalayas contain two species; at least two or three others are found in the Chinese empire; and to the flora of Japan six species are credited. One of the Japanese species, however, Carpinus erosa of Blume, is a doubtful plant; another, the Carpinus Tschonoskii of Maximowicz, from the Hakone Mountains and the region of Fuji-san, I have never seen; and a third, Carpinus Yedoensis, a small tree cultivated in gardens in the neighborhood of Tōkyō, is, perhaps, like many of the plants cultivated by the Japanese, a native of China. Three species are certainly indigenous to the Japanese soil.

Carpinus laxiflora resembles, in the character of the bark, in the size and shape of the leaves, and in the structure of the flowers, the European and American Hornbeams. It is a graceful tree, occasionally fifty feet in height, with a trunk eighteen to twenty inches in diameter, covered with smooth pale, sometimes almost white, bark, and slender branches. The leaves are ovate or ovate-elliptical, rounded or subcordate at the base, contracted at the apex into long slender points, and doubly serrate; they are dark green above, pale yellow-green below, three to four inches long, an inch to an inch and a half broad, and prominently many-veined, and in the autumn turn yellow or red and yellow. The fruit is produced in lax hairy catkins four or five inches long, with spreading oblique prominently veined bracts, which are obscurely lobed, more or less infolded at the base round the fruit, and nearly an inch long. This fine tree is common in all the mountain forests of Hondo, where it is most abundant at elevations between two and three thousand feet above the sea; in Yezo it reaches the southern shores of Volcano Bay, where, near the town of Mori, it is common in Oak forests, and grows to its largest size.

The other Japanese species of Carpinus differ from Carpinus laxiflora and from the American and European species in their furrowed scaly bark, in the stalked bract of the male flower, in the closely imbricated bracts of the fruiting catkins, which look like the fruit of the Hop-vine, and in the form of these bracts, which are furnished at the base with a lobe which covers the fruit, and is more or less inclosed by the infolding of the opposite side of the bract. On account of these differences these two trees are sometimes referred to the genus Distigocarpus, founded by Siebold & Zuccarini to receive their Distigocarpus Carpinus.

The figure (see Plate xxi.) shows flowering and fruiting branches of this tree, a staminate flower, and a bract of the fruiting catkin. Botanists now pretty generally agree that the characters upon which Distigocarpus was founded are not of sufficient importance to justify its separation from Carpinus; and Distigocarpus Carpinus, if the oldest specific name is used, becomes Carpinus Carpinus. By Blume, who first united Distigocarpus with Carpinus, it was called Carpinus Japonica, the name under which it has appeared in all recent works on the Japanese flora. It is a tree forty to fifty feet in height, with a trunk often twelve to eighteen inches in diameter, and wide-spreading branches which form a broad handsome head. The branches are slender, terete, and coated at first with long pale hairs, and later are covered with dark red-brown bark often marked with oblong pale lenticels. The winter-buds are half an inch long, acute, and covered with many imbricated thin light brown papery scales; with the exception of those of the outer rows they are accrescent on the growing shoots, and at

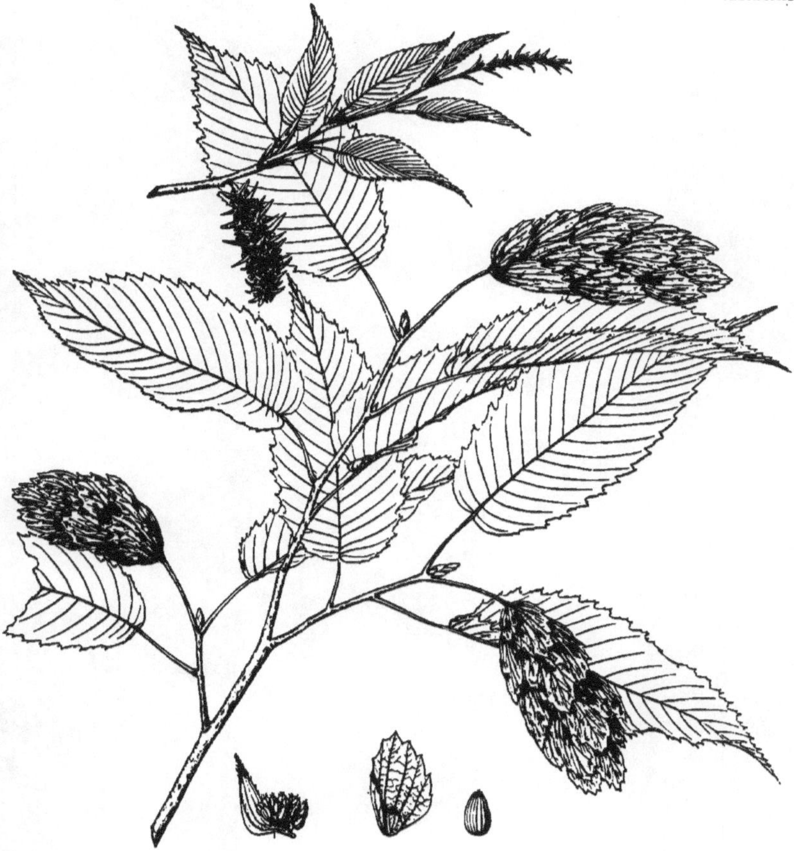

C. E. Faxon del.

CARPINUS CARPINUS, Sarg.

maturity are nearly an inch long and hairy on the margins. The leaves are ovate, long-pointed, slightly and usually obliquely cordate at the base, coarsely and doubly serrate, thick and firm, dark green on the upper surface, paler on the lower surface, three or four inches long and about an inch and a half wide, with stout midribs and many straight prominent veins slightly hairy below and deeply impressed above. The stipules are linear, acute, scarious, an inch long, and covered with pale hairs. The male inflorescence is an inch long, with stalked lanceolate-acute bracts half an inch in length, and more or less ciliate on the margins. The female inflorescence is two thirds of an inch long, and is raised on a slender stem coated, like the bracts which subtend the ovaries, with thick white tomentum; the outer bracts are acute, scarious, a quarter of an inch long, and early deciduous; the inner bracts are oblique, coarsely serrate toward the apex, conspicuously many-ribbed, and furnished at the base with a minute ovate serrate lobe which covers the ovary. Before the fruit ripens these inner bracts enlarge until they are two thirds of an inch long and one third of an inch broad, and are closely imbricated into a cone-like catkin which resembles in shape, color, and texture that of our American Hop Hornbeam; it is, however, often two or two and a half inches long. The nutlet is slightly flattened, with about ten straight prominent ridges extending from one end to the other.

Carpinus Carpinus is common on the Hakone and Nikkō mountains between two and three thousand feet elevation above the sea; it apparently does not range very far north in Hondo or reach the island of Yezo. This interesting and beautiful tree, which is remarkable among Hornbeams in the character of the bark and in the female inflorescence, appears to be perfectly hardy in New England. For a number of years it has inhabited the Arnold Arboretum, and during the last two seasons has produced flowers and fruit here. In its young state it makes a handsome, compact, pyramidal, bushy, and very distinct-looking tree.

But the most beautiful of the Hornbeams of Japan, as it appears in the forests of Yezo, is Carpinus cordata, which often attains the height of forty feet, with a stout trunk sometimes eighteen inches in diameter, covered with dark deeply furrowed scaly bark. The stout branchlets are orange-color, or light brown when they are three or four years old, and are covered with large oblong pale lenticels. This species is remarkable in the size of its winter-buds, which are fully grown by midsummer, and are sometimes nearly an inch in length, and acute, and covered with light chestnut-brown papery scales. The leaves are thin, broadly ovate, pointed, deeply cordate, doubly serrate, six or seven inches long, and three or four inches broad; they are light green on both surfaces, although rather lighter colored on the lower, with conspicuous yellow midribs and veins slightly hairy below and impressed above. The catkins of fruit are often five or six inches long and an inch and a half wide, with broadly ovate, remotely serrate bracts; their basal lobe is proportionately much larger than that of the last species, and is sometimes a third of the length of the bract, to which it is often united along nearly its entire length, while in Carpinus Carpinus the lobe is only attached at the base.

This is the only species of central Yezo, where it is one of the common forest-trees, growing with Oaks, Magnolias, Ashes, Walnuts, Acanthopanax, Birches, etc.; it also grows in Hondo at high elevations, although it is here much less common than farther north. This fine tree is apparently still unknown in American and European gardens; it is one of the largest of the Hornbeams, and certainly one of the most distinct and beautiful of them all. As it grows in its native forests with a number of trees which flourish here in New England, it may be expected to grace our plantations with its stately habit, large leaves, and long clusters of fruit.

Ostrya, the Hop Hornbeam, appears in North America with two species, one, a small forest-tree of the eastern states, the second known only in the Grand Cañon of the Colorado in Arizona; a third species inhabits southern Europe, Asia Minor, and the Caucasus; the genus has no representative in the Himalaya forest region; and, so far as I know, has not been found within the borders of the Chinese empire or in Manchuria. It appears again in northern Japan, however, where the Hop Hornbeam is one of the rarest of Yezo trees. Maximowicz, who found it in the southern part of that island, considered the Japanese Ostrya a variety of our American species and called it Ostrya Virginica, var. Japonica.[1] The American and the Japanese trees are very similar in botanical characters; indeed, it is difficult to find characters to separate satisfactorily the species of this genus, which might all be considered geographical varieties of one. The Japanese and American trees, however, look very differently in the forest, and there are differences in the bark which are not easy to express in words. The leaves of the Japanese tree are thinner and the heads of fruit are smaller than those on the American species (see Plate xxii.). Unfortunately, I have not had an opportunity to examine the flowers of the Japanese tree; it is not probable, however, that they would afford a character by which the species could be distinguished. In the forests of Yezo I felt no doubt of its specific distinctness; the meagre and unsatisfactory material of the herbarium rather shakes than confirms this opinion. But, all things considered, it is, perhaps, best to consider the Japanese tree as specifically distinct. Not until it has been grown here during many years side by side with the American species will it be possible to reach any opinion on this subject worthy of much consideration. If it proves to be distinct it should bear the name of Ostrya Japonica. In the neighborhood of Sapporo the Japanese Ostrya is rare; here in low moist woods, growing with Oaks, Acanthopanax, and Aralia, it sometimes attains a height of eighty feet, and forms a tall straight trunk eighteen inches in diameter. We saw only one such tree, in the grounds attached to the headquarters of the Forest Department of Hokkaido, and only two or three other individuals; these were much smaller, perhaps not more than twenty feet high, and were scattered over the Sapporo hills. We saw nothing of this tree in southern Yezo or in northern Hondo, where Tschonoski, Maximowicz's servant and collector, found it on the high mountains of the province of Nambu.

[1] *Mél. Biol.* xi. 317.

PLATE XXII.

C. E. Faxon del.

OSTRYA JAPONICA, Sarg.

THE OAKS, CHESTNUTS, WILLOWS, AND POPLARS.

ALTHOUGH poorer in species and less important in the number, size, and value of individuals than in eastern America, Quercus furnishes one of the principal elements of the forests of Japan. The types are all of the Old World, and there is nothing in Japan which corresponds with our Red, Black, or Scarlet Oaks, or with the Black Jack, the Willow Oak, the Shingle Oak, the Turkey Oak, the Spanish Oak, the Water Oak, or the Pin Oak, the Blue Jack, or with our Chestnut Oaks. In the north and on the high mountains of Hondo there are four White Oaks, and in the south a number of species with evergreen foliage of sections of the genus, which are not represented in the United States. In the south, too, there are a couple of deciduous-leaved species with biennial fructification of the Turkey Oak (Quercus Cerris) sort.

The best known of the Japanese Oaks to European and American planters is Quercus dentata (the Quercus Daimio of gardens). This tree is remarkable for the great size of its leaves, which are often a foot long and eight inches broad, obovate in outline, and deeply serrately lobed, and for the long loose narrow chestnut-brown scales of the large cup which nearly incloses the small-pointed acorn. In central Hondo this tree is found only on the high mountains, and it is not at all common; but in the extreme northern part of the island it appears in great numbers on dry gravelly slopes, at no great elevation above the sea. Here, apparently, however, it does not reach the size it attains farther north, and the finest trees we saw were on the gravelly plain south of Volcano Bay, and in the neighborhood of Sapporo. The illustration (see Plate xxiii.) represents a group of these trees growing just outside of Sapporo, and shows their habit at maturity. Although Quercus dentata grows to the height of at least eighty feet, and forms a thick trunk more than three feet in diameter, it is not an imposing or handsome tree in its maturity, and is only beautiful in youth. Old trees lack symmetry and the appearance of strength, and are sprawling in habit, without being picturesque. The bark is rather dark for a White Oak, and not unlike that of our Rock Chestnut Oak (Quercus Prinus); it is valued for tanning leather, but the wood is considered worthless. Quercus dentata appears to be the only deciduous-leaved Oak cultivated by the Japanese, and small trees are common in the gardens of Tōkyō and other southern cities, where, however, they seem to languish. A variety (pinnatifida), with deeply divided leaves, is cultivated in the Botanic Garden at Tōkyō, and has, I believe, been introduced into Europe.

In central Yezo two noble White Oaks, Quercus crispula and Quercus grosseserrata, form a considerable part of the forest-growth. The Dutch botanist Miquel considered them forms of one species; but Professor Miyabe, who has had the best opportunity for studying these trees under the most favorable conditions, believes them to be distinct in their fruit, although similar in foliage. In Quercus crispula he finds "the cup deeper, embracing about half the cylindrical nut, falling off with it when ripe; while in the latter, Quercus grosseserrata, the cup is hemispherical, inclosing about a third of the oblong ovoid nut, which falls off free when ripe." His view, too, that Quercus grosseserrata cannot be specifically distinguished from the Saghalin and Manchurian Quercus Mongolica, will probably be found to be correct. Quercus crispula appears to range farther south than Quercus grosseserrata, which extends north to the Kurile Islands, and was not recognized by us in Hondo; on the Nikkō Mountains, on the road to Lake Chuzenji, we saw, however, fine forests of Quercus crispula. In central Yezo, where the two species grow side by side on the hills, Quercus crispula appears the more common tree on low ground, near the banks of streams. Both have elliptical or obovate-oblong coarsely and irregularly lobed leaves, resembling in color and texture those of the common Oak of Europe. Their bark is pale or sometimes dark, and scaly; and both species under favorable conditions rise to a height of eighty to a hundred feet, and produce stems three to four feet in diameter. Both are timber-trees of the first class, and both, should they thrive in this country, may be expected to add beauty and interest to our parks and plantations. The smaller, shorter acorn of Quercus crispula appears to offer the only character for distinguishing the two trees; in their port, bark, and foliage they were indistinguishable to my eyes.

The fourth Japanese White Oak, Quercus glandulifera, ranges in Yezo nearly as far north as Sapporo, although it is only south of Volcano Bay that it is really abundant. This, the common Oak of the high mountains of central Japan at elevations over 3,000 feet, is probably the most widely distributed species of the empire; it is a pretty tree, rarely more than thirty or forty feet high, although on the hills above Fukushima, on the Nakasendō, we saw specimens nearly twice that height. The leaves are narrowly obovate or lanceolate-acute, glandular-serrate, pale or nearly white on the lower surface, and from one to four inches in length. The acorns are small, acute, and inclosed at the base only by the shallow thin-walled cups covered with minute appressed scales. Like many American Oaks, this species varies remarkably in the size of individuals, and in some parts of the country traversed by the Nakasendō we found plants only a foot high covered with acorns. This Oak was sent to the Arnold Arboretum many years ago from Segrez by Monsieur Lavallée. It is perfectly hardy here, and has flowered for years, although it remains a bush, making no attempt to grow into a tree.

Of the other deciduous-leaved Oaks, Quercus serrata, one of the most widely distributed of the Asiatic species, ranging, as it does, from Japan to the Indian Himalaya, is common in dry soil near the coast below Yokohama and on the foothills of the mountains of central Hondo. It is a small tree, twenty to forty feet high, with a slender black-barked trunk and beautiful dark green lustrous oblong acute leaves, their coarse teeth ending in long slender mucros, and with small acorns inclosed in cups covered with long, loose, twisted, and reflexed scales coated with soft pale tomentum. In Japan this tree appears to spring up in waste lands in great numbers; it is only valued for the charcoal which is made from it.

Quercus variabilis, a nobler tree of the same general character, we saw only in the grounds of a temple near Nakatsu-gawa, on the Nakasendō, where there were specimens fully eighty feet high, with tall straight trunks three or four feet in diameter, covered with thick pale corky bark, which is sometimes used by the Japanese for the same purposes that we use the

PLATE XXIII.

QUERCUS DENTATA, Thunb.

THE OAK FAMILY.

bark of the Cork Oak. The leaves are oblong-oval, pointed, less coarsely toothed than those of Quercus serrata, dark green and lustrous above, and pale, or nearly white, below. From Quercus serrata, too, it differs in the smaller cups and in their shorter thicker scales. A number of plants have been raised in the Arboretum from the acorns which we picked up under these trees, and if they are not hardy here in New England they will certainly thrive in the middle states.

It is impossible to know whether many of the evergreen Oaks which we saw in Japan were growing naturally or had been planted. In the gardens and temple grounds of Tōkyō, Yokohama, Kyōto, and other southern cities evergreen Oaks are the commonest trees; but we did not see them growing in the forest except near temples. The species most frequently seen in Tōkyō and Yokohama are Quercus cuspidata and Quercus glauca; they are both large and beautiful trees, said to be particularly conspicuous in early spring from the bright red color of their young shoots and new leaves, which at that season make a charming contrast with the dark and lustrous green of the older foliage. They should be introduced into our southern states, where, probably, all the Japanese evergreen Oaks will flourish. The wood produced by Quercus cuspidata and Quercus glauca does not appear to be valued in Japan, but the acorns of the latter are of considerable commercial importance, and are cooked and eaten by the Japanese.

Quercus acuta, which is also much planted in Tōkyō, we saw growing to the height of more than eighty feet, with Quercus variabilis, in the temple grounds at Nakatsu-gawa, and also near the temple of Higane, near Atami, on the coast. It is a noble tree, with ovate, acute, long-pointed, dark green, thick, and lustrous leaves. Quercus acuta has been introduced into English gardens, with a number of other evergreen Japanese Oaks, through the efforts of the Veitches, who obtained it some years ago from their collector, Maries. But the finest Oak-tree, and perhaps the finest tree which we saw in Japan, was a specimen of Quercus gilva in the temple grounds at Nara, where there are a number of remarkable specimens of this beautiful species, which is distinguished by its lanceolate-acute leaves, which are glandular-serrate only above the middle, bright green on the upper, and thickly coated on the lower surface, like the young branches, with pale or slightly ferrugineous tomentum. The largest of these Nara trees was probably a hundred feet high, with a trunk covered with pale scaly bark, which, breast-high from the ground, girthed just over twenty-one feet; it rose without a branch, and with little diminution of diameter, for something like fifty feet, and then separated into a number of stout horizontal branches, which had not grown to a great length, and formed a narrow cylindrical round-topped head.

Of the other Japanese Oaks, Quercus Thalassica, Quercus Vibrayiana, and Quercus glabra, we only saw occasional plants in gardens. The Quercus lacera of Blume we did not see at all.

The Chestnut-tree is widely distributed through the mountain forests of Japan, and seems to have received some attention as a fruit-tree from the Japanese, who recognize a number of large-fruited varieties. Very large chestnuts appear in profusion in the markets of Aomori, and are said to be produced in the immediate neighborhood of that northern town. But the largest chestnuts of Japan, which equal in size the best marrons of southern Europe, are found in the markets of Kobe and Osaka. It is these Kobe marrons which are now sent to San Francisco in considerable quantities. Rein, whose book on the industries of Japan contains the fullest and most exact account of Japanese rural economy which has yet been written, believed that the chestnut was less used in Japan as an article of human food than in Europe, but I have never seen chestnuts offered in such quantities in the markets of any American or European city as in those of Tōkyō and other Japanese towns. The Chestnut-trees which we saw had the appearance of growing spontaneously; and we saw nothing like an orchard of these trees, which, so far as we were able to observe, are not planted near dwellings or temples or for shade. In Japan the Chestnut-tree grows at least as far north as central Yezo, and is scattered through the mountain forests of Hondo, where it is most abundant at elevations of about 2,500 feet above the ocean, growing on steep slopes in small open groves or mixed with trees of other kinds. We saw no evidences that the Chestnut-tree grows in Japan to the noble dimensions it sometimes reaches in Europe and on the slopes of the southern Alleghany Mountains, and specimens over thirty feet high, with trunks more than a foot in diameter, were rare in that part of the country which we visited. The Japanese Chestnut appears to be more precocious than the American tree, and saplings only ten or twelve feet high are often covered with fruit. The large-fruited northern form from the neighborhood of Aomori should be brought to this country, as it may be expected to support a greater degree of cold than the French or Kobe Marrons, and, therefore, to be available for cultivation much farther north here. By its introduction it is possible that marron-growing may become a profitable industry in states with climates as severe as those of Wisconsin, Michigan, and Massachusetts.

The Beech in Japan is one of the noblest trees of the forest, as it is in eastern North America and in Europe. Its range is similar to that of the Horse-chestnut, in the north appearing on the shores of Volcano Bay in Yezo only a few feet above the level of the ocean, and extending southward along the mountains of the other islands. It is, perhaps, the commonest deciduous tree of the mountains of Hondo, where, between 3,000 and 4,000 feet, or toward the upper limits of the deciduous forest, it sometimes covers wide areas, nearly to the exclusion of other trees, or sometimes grows mixed with Oaks, Chestnuts, and occasional Firs and Spruces. Trees eighty or ninety feet tall, with trunks more than three feet in diameter, are not uncommon. The fact that Beech-wood is little used by the Japanese, and the comparatively inaccessible situations where it is mostly found, account, no doubt, for the abundance of this tree in Japan and the existence of so many large individuals. This, the Asiatic form of the European Fagus sylvatica, is hardly to be distinguished from the European tree, which it resembles in every essential character. The variety Sieboldii (the Fagus Sieboldii of Endlicher) I looked for in vain, and I hazard the opinion that it will turn out to be a tree of the herbarium and not of the forest.

I can throw no light upon the Japanese Willows which abound at the north in numerous continental mostly shrubby forms; they require more careful investigation than it was possible to give them during our hurried autumnal visit, when the flowers and fruit had disappeared. On the streets of Europeanized Tōkyō, Willows are now chiefly planted as shade-trees; they are the Weeping Willow (Salix Babylonica), an inhabitant of China, and a favorite with the Japanese, and Salix eriocarpa of Franchet & Savatier, a species which we saw growing by river-banks on the Nakasendō, and which looks too much like Salix alba to be distinct from that species which might be expected to reach Japan. But the handsomest Willow we saw in Japan, and certainly one of the most beautiful of all Willows, is Salix subfragilis, which appears to be confined to Japan, where it was discovered in the neighborhood of Hakodate by Charles Wright. We were first struck by the beauty of this tree between Nikkō and Lake Chuzenji, where there are a few specimens on the banks of the mountain torrent which the road follows in ascending the mountains. It was at Sapporo, however, that this Willow appeared in its greatest beauty; here on the banks of streams Salix subfragilis forms trees at least fifty feet in height, with short stout trunks three or four feet in diameter, covered with thick deeply furrowed bark, and stout branches which spread nearly at right angles, like those of an old pasture Oak. The leaves are oblong, acute, rounded at the base, and coarsely crenulate-serrate; they are borne on stems an inch and a half in length, and are six or seven inches long, two or two and a half inches broad, dark green and lustrous on the upper surface, and silvery white on the lower; the stipules are foliaceous, obliquely rounded, and rather more than half an inch across. This Willow appears to be one of the most desirable trees to introduce into our collections, and the only Japanese Willow we saw of real value, from a horticultural point of view.

Populus is poorly represented in Japan; the two species which are found in the empire are both of Old World types, and there is nothing which corresponds to the Cottonwoods, which line the river-banks in all the central and western regions of this continent. The Aspen of Europe appears in one of its forms in Japan (Populus tremula, var. villosa), looking, however, so distinct from the continental Aspen that it is hard to believe that it is not specifically distinct; it is the Populus Sieboldii of Miquel, the oldest name. This tree is not rare in southern Yezo, where it grows to the height of twenty or thirty feet, springing up in considerable numbers on dry, gravelly soil. We saw it in the greatest perfection on the plains south of Mori, on Volcano Bay, and less commonly on the mountains near Aomori in Hondo. Of the second species, the Populus suaveolens of Fisher, we encountered a few individuals in southern Yezo, where it is probably near the southern limit of its range, it being a tree of Saghalin and the Amour country. It is evidently only a form of the Balsam Poplar, which is found in all northern regions, where, especially in some parts of British America, it constitutes by far the largest part of the forest-growth. In Japan the Balsam Poplar grows to an immense size, and some individuals which we saw were certainly eighty and perhaps a hundred feet tall, with long trunks five or six feet in diameter, rising like sentinels above the low mostly second-growth forests of southern Hokkaido.

THE CONIFERS.

In cone-bearing plants Japan is somewhat richer than eastern America. All of our genera, with the exception of Taxodium, the Bald Cypress of the southern states, are represented in the empire, where two endemic genera occur, Cryptomeria and Sciadopytis, and where Cephalotaxus and, perhaps, Cupressus have representatives. In cone-bearing species, too, owing to the greater multiplication of forms of Abies, Japan is richer than eastern America, where we have only two indigenous Firs. The genus Pinus, which furnishes a very considerable part of the forest-growth on the Atlantic seacoast, and which is represented here by thirteen species, has only five in Japan; and of these two are small trees of high altitudes and one is an alpine shrub. Japan is richer than eastern America in Spruces, of which we have only two, in Chamæcyparis, and in Juniperus. The two floras each contain a single Thuya, a Taxus, a Tumion, two Hemlocks, and a Larch. In Japan, Conifers are more planted for shade and ornament than they are in America, or, perhaps, in any other country, although, except above 5,000 feet in Hondo, where there are continuous forests of Hemlock, they form a small part of the composition of indigenous forest-growth; and forests of Pines, Spruces, or Firs, such striking features in many parts of this country, do not occur, except, perhaps, in northern Yezo, which we did not visit, and where there are said to be great forests of Abies Sachalinensis.

The Japanese Arbor-vitæ, Thuya Japonica, which is sometimes found in our plantations under the name of Thuyopsis Standishii, and which is more like the species of the northwest coast (Thuya gigantea) than our eastern Arbor-vitæ or Yellow Cedar, appears to be a rare tree in Japan, and we saw only a few solitary individuals on the shores of Lake Chuzenji and of Lake Yumoto in the Nikkō Mountains. Here it was a formal pyramidal tree twenty or thirty feet high, with pale green foliage and bright red bark.

Thuyopsis dolobrata, which is, perhaps, best considered a Cupressus rather than a Thuya, is a tree of high altitudes. In the Nikkō Mountains above Lake Yumoto it is common between 5,000 and 6,000 feet above the sea-level, growing as an under-shrub under the shade of dense Hemlock forests, and here, in favorable positions, sometimes finally rising to the height of forty or fifty feet, with a slender trunk covered with bright red bark, long pendulous graceful lower branches, and a narrow pyramidal top. This handsome tree finds its northern home on the mountains which surround the Bay of Aomori, in northern Hondo. It is a species which evidently requires shade, at least while young, and even the older plants, where we saw them, were always surrounded and overtopped by taller trees. The elevation at which this tree grows indicates that it should prove hardy here if properly protected from the sun, especially during the winter, for in Japan the young plants are not only shaded by the coniferous forest above them, but are buried during several months under a continuous covering of snow. Under proper conditions this tree will be found to be one of the best plants to form undergrowth in coniferous forests. The wood is considered valuable, and, owing to its durability, is used in boat and bridge building. We saw planted trees in the coniferous forests on the mountain-slopes near Nakatsu-gawa, in the valley of the Kisogawa, but no other indication that it is valued as a timber or ornamental tree by the Japanese, who, according to Dupont,[1] have produced a number of varieties, of which the one with variegated foliage only has reached our gardens.

Of all the Japanese Conifers the most valuable is the Hi-no-ki, Chamæcyparis (Retinospora) obtusa. In the forests planted on the lower slopes of mountains in the interior of Hondo, and in some of the temple groves, notably in those of Nikkō, this fine tree attains a height of a hundred feet, with a straight trunk without branches for fifty or sixty feet, and three feet through at the ground. At elevations between 2,000 and 3,000 feet above the sea, usually on northern slopes and in granitic soil, which it seems to prefer, the Hi-no-ki is largely planted as a timber-tree; indeed, only the Cryptomeria, which seems to be less particular about soil and exposure, is more planted for timber in Japan. The tree is sacred among the disciples of the Shintō faith, and is, therefore, cultivated in the neighborhood of all Shintō temples, which are built exclusively from Hi-no-ki wood. The palaces of the Mikado in Kyōto were always made of it, and the roof was covered with long strips of the bark. It is considered the best wood to lacquer; at festivals food and drink are offered to the gods on an unlacquered table of this wood, and the victim of harakari received the dagger upon a table of the same material. It is used for the frames of Buddhist temples and for the interior of the most carefully finished and expensive houses. The wood is white or straw-color, or sometimes pink, and in grade, texture, and perfume resembles that of the Alaska Cedar, Chamæcyparis Nootkatensis. Like the wood of that tree, it has a beautiful lustrous surface, and is straight-grained, light, strong, and tough, and remarkably free from knots and resin. In America we have no wood of its class which equals it in value, with the exception of that furnished by the two species of Chamæcyparis of the Pacific coast, and the Hi-no-ki might be introduced with advantage as a timber-tree into those parts of the eastern states where it could find conditions which would insure its growth. It has proved perfectly hardy in this country as far north as Massachusetts, but the sea-level or the dry summers here do not suit any of the Retinosporas, which give no promise of long life or great usefulness anywhere on the Atlantic seaboard. They should be tried, however, on the slopes of the southern Alleghanies where they could find conditions not very unlike those in which they flourish in their native land.

The second species of this genus, Chamæcyparis pisifera, the Sawara, is a less valuable tree than the Hi-no-ki, although the two species are always found growing together in plantations and in temple gardens; indeed, they can only be distinguished after some practice, unless the cones are examined, although after a few days among them the more ragged crown, with its looser and more upright branches, of the Sawara stands out clearly to the eye in contrast with the Hi-no-ki with its rounder top and more pendulous branchlets. The wood is of a reddish color, of a rougher grain, and less valuable than that of the Hi-no-ki, although the two trees are planted in about equal numbers. As it grows here in our gardens, Chamæcyparis pisifera is a less ornamental plant than Chamæcyparis obtusa; it grows, however, more vigorously, and

[1] *Les Essences Forestières du Japon.*

promises to live longer and attain a greater size. All the other Retinosporas of our gardens are juvenile or monstrous forms of these two trees. Some of the dwarf forms are much cultivated in Japan especially as pot-plants, but they are not as popular there as I had been led to expect, and are most often seen in the nursery-gardens of the treaty ports, where they are collected to please the fancy of foreign purchasers.

The most generally planted timber-tree of Japan is the Sugi, Cryptomeria Japonica, and its wood is more universally used throughout the empire than that of any other Conifer. It is one of the common trees of temple gardens and roadside plantations, and, when seen at its best, as in the temple groves of Nikkō or Nara, where it rises to the height of a hundred or a hundred and twenty-five feet, with a tall shaft-like stem tapering abruptly from a broad base, covered with bright cinnamon-red bark and crowned with a regular conical dark green head, it is a beautiful and stately tree which has no rival except in the Sequoias of California. Great planted forests of the Cryptomeria appear all over Hondo on broken foothills and mountain-slopes up to elevations of nearly 3,000 feet above the sea, low valleys and good soil being usually selected for such plantations, as the trees need protection from high winds. The plantations decrease in size and luxuriance in northern Hondo, and the cultivation of the Sugi does not appear to be attempted north of Hakodate, where there is a grove of small trees on the slope of the hill above the town. The wood is coarse-grained, with thick layers of annual growth, dark reddish heartwood, and thick pale sapwood; it is easily worked, strong and durable, and is employed in all sorts of construction. The bark, which is carefully stripped from the trees when they are cut down, is an important article of commerce, and is used to cover the roofs of houses. A large round bunch of branchlets covered with their leaves hung over the door of a shop is the familiar sign of the dealer in saké.

Japan owes much of the beauty of its groves and gardens to the Cryptomeria. Nowhere is there a more solemn and impressive group of trees than that which surrounds the temples and tombs at Nikkō, and the long avenue of this tree, under which the descendants of Ieyasu traveled from the capital of the Shoguns to do honor to the burial-place of the founder of the Tokugawa dynasty, has not its equal in stately grandeur. This avenue, if the story told of its origin is true, can teach a useful lesson, and carries hope to the heart of the planter of trees, who will see in it a monument more lasting than those which men sometimes erect in stone or bronze in the effort to perpetuate the memory of their greatness. When the body of Ieyasu was laid in its last resting-place on the Nikkō hills, his successor in the Shogunate called upon the Daimyos of the empire to send each a stone or a bronze lantern to decorate the grounds about the mortuary temples. All complied with the order but one man, who, too poor to send a lantern, offered instead to plant trees beside the road, that visitors to the tomb might be protected from the heat of the sun. The offer was fortunately accepted, and so well was the work done that the poor man's offering surpasses in value a thousand-fold those of all his less fortunate contemporaries.

Something of the stateliness of this avenue appears in our illustration (Plate xxiv.) although, without the aid of colors, it is impossible to give an idea of the beauty of the Cryptomeria. The planted avenue extends practically all the way from Tōkyō to Nikkō, but it is only when the road reaches the foothills that it passes between rows of Cryptomerias, the

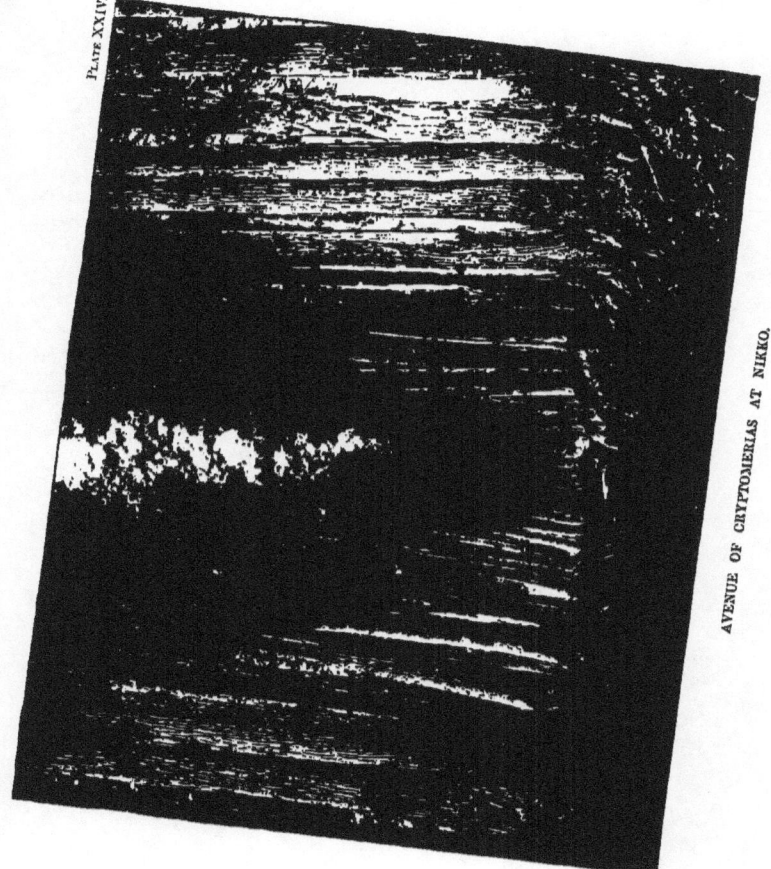

AVENUE OF CRYPTOMERIAS AT NIKKO.

lower part being planted, as is the case with the other great highways of Japan, with Pine-trees; nor is this avenue continuous, as has often been stated, for whenever a village occurs or one of the roadside tea-houses, which are scattered all along the road, there is a break in the rows of trees, and it is only in some particular spots that a long view of continuous trees is obtained. The railroad, which follows parallel and close to the avenue for a considerable distance and then crosses it just before the Nikkō station is reached, is a serious injury to it. The trees, as will be seen in the illustration, are planted on high banks made by throwing up the surface-soil from the roadway; they are usually planted in double rows, and often so close together that sometimes two or three trees have become united by a process of natural grafting. Young trees are constantly put in to fill gaps, and every care apparently is taken to preserve and protect the avenue. How many of the trees originally planted when the road was first laid out in the beginning of the seventeenth century are left it is impossible to say, but I suspect that most of those now standing are of much later date. One of the trees close to the upper end of the road which had been injured by fire was cut down during our visit to Nikkō. The stump, breast-high above the ground, measured four feet across inside the bark, and showed only one hundred and five layers of annual growth. Few of the trees in the avenue were much larger than this, although in the neighborhood of the temples there are a few which girt over twenty feet; these were probably planted when the grounds were first laid out.

The two, Chamæcyparis and the Cryptomeria, the most valuable timber-trees in Japan, are now almost unknown in a wild state. They may, perhaps, be found growing naturally on some of the southern mountains which we did not visit; but wherever we went, we saw only trees that had been planted by man, although some of the plantations had evidently lived through several centuries.

Cephalotaxus drupacea is the only Japanese member and the type of a genus of half a dozen species distributed from Japan, through China to northern India. It is widely and quite generally scattered through the mountain regions of the empire, extending north to central Yezo, where it appears on the low hills as an under-shrub only two or three feet high, while on the Hakone Mountains, in Hondo, it occasionally grows into a bushy tree twenty or twenty-five feet in height. Cephalotaxus drupacea is now common in our gardens, although it is not very hardy or satisfactory here in New England, where it often suffers in winter, missing, no doubt, the thick and continuous covering of snow which protects it in Yezo. Like its relative, the Gingko, the same individual does not produce male and female flowers, and the fruit, like that of the Gingko, is an almond-like nut inclosed in a fleshy covering. A resinous oil, used in lamps, is pressed from the seeds, and the wood is occasionally employed in cabinet-making.

The Gingko, although we are in the habit of associating it with Japan, is in reality not a native of that country, into which it was brought with their religion by the Buddhist priests. It is still unknown in a wild state, and it is possible that this genus, which was widely distributed with many species through the temperate and colder parts of the northern hemisphere in tertiary times, has become exterminated from its native forests, and has only been preserved through the agency of the priests of Buddha, who seem to hold it in particular respect. The hardiness of this beautiful tree, which thrives under the most trying conditions and in the severest climates, indicates that it originated in some northern interior region; and if it is ever seen growing without cultivation it will probably be in some remote district of Mongolia. There are noble great broad-branched specimens in the neighborhood of the temples in Tōkyō fully a hundred feet high, with tall massive trunks six or seven feet in diameter. The Gingko is, perhaps, the most beautiful, as it is certainly the most interesting tree which is to be seen in Japan; and in the autumn, especially when the sunlight flutters through the bright yellow leaves, these great trees, with their broad heads of graceful semipendulous branches, are magnificent objects. The fleshy covering of the fruit has a rancid and most disagreeable flavor, but the kernel of the almond-like stone is delicate and is esteemed a luxury in both China and Japan, where it is found in the markets in considerable quantity. The wood, which is light yellow in color, is soft and brittle, and as the trees grow to a very great age and are planted only for ornament in Japan and rarely cut down, it has no economic importance there.

Torreya, or, if the custom which now prevails among American botanists is followed, Tumion, Rafinesque's name, which also appears in eastern and western America and in China, occurs in Japan in its largest and most beautiful representative, Tumion nuciferum, one of the handsomest of all coniferous trees. Although nowhere very common, the Kaya, as this tree is called in Japan, was seen in all the mountainous regions of central Hondo which we visited. It often grows as an under-shrub in the forest, or as a small tree twenty or thirty feet tall, but occasionally rises to the dignity of a tree of the first class, as on the banks of the Kisogawa, near Agematsu, where we saw specimens fully eighty feet high, with great trunks four or five feet in diameter. Such trees, with their bright red bark and compact heads of dark green, almost black lustrous foliage, possess extraordinary beauty. No other Yew-like tree which I have seen equals it in massiveness and depth of color, and the Kaya should be cultivated wherever the climate permits it to display its beauty. The elevation above the sea at which it flourishes in Japan indicates that it will be hardy in the middle states, although we cannot expect to see it grow to any size in New England. An oil used in cooking, kaya-no-abura, is an article of considerable commerce in Japan, and the kernels of the nuts, which possess an agreeable, slightly resinous flavor, are sold in great quantities in the markets in the autumn, and are a favorite article of food. The wood is strong, straight-grained, light yellow, and valued in building and cabinet-making.

Taxus, which has two species in eastern America, one in the north and another, almost the rarest of American trees, in the south, which is represented in western North America, and is widely distributed through Europe and continental Asia, appears in Japan with a noble tree, Taxus cuspidata, which, to judge by our observations, is confined to the island of Yezo, where it is not uncommon on the low hills of the interior. Here it often attains the height of forty or fifty feet, and forms a trunk two feet in diameter, covered with bright red bark. The Yew is often employed by the Japanese to ornament their gardens, and the wood, which is exceedingly hard, tough, and of a bright red color, is used by the Ainos for their bows, and is valued in cabinet-making and for the interior decorations of expensive houses. This beautiful tree, as is now well known, flourishes in this country, where it has proved itself perfectly hardy, promising to be really valuable as an ornamental plant.

It is not thought now that Podocarpus, a genus of the tropical and subtropical regions of both hemispheres and of Tasmania, is indigenous in Japan, although two species are often cultivated there. The more common is Podocarpus macrophylla, a small tree with lanceolate acute leaves, and a common hedge-plant in Tōkyō gardens, in which it is often cut into fantastic shapes. It is a much less beautiful, although a hardier, tree than Podocarpus Nageia, with its thick broad glossy leaves and beautiful purple trunks, the second species seen in Japan. It is one of the favorite subjects, especially in a variety in which the leaves are marked by broad white stripes, for dwarfing and pot-culture. The real beauty of this tree is only seen, however, when it has become large and old and the trunk is covered with its peculiar smooth purple bark. A grove of these trees on the hill behind the Shintō temples at Nara is one of the most interesting spots in Japan, and in solemn dignity and beauty is only surpassed by the grove of Cryptomerias which surrounds the mausoleums of Ieyasu and Iemitsu at Nikkō.

Like Cryptomeria, Sciadopitys is monotypic and endemic to Japan. It is one of the most curious and interesting of trees, with scale-like leaves in whose axils are produced the phylloid shoots, which are generally mistaken for the leaves, and which are arranged near the ends of the branches like the ribs of an umbrella, — a peculiarity to which this tree owes its familiar English name, the Umbrella Pine. Like the Gingko, the Sciadopitys was for a long time known only from a few individuals cultivated in temple gardens and from the grove on the hill in Kyūshū, where the ancient monastery town of Kôya stands, to which the Sciadopitys owes its Japanese name, Kôya-maki. There is said to be a remarkable grove of these trees here, which was once supposed to be the original home of the species; but Rein and other writers now agree in thinking that they were originally planted by the monks. Dupont found what he considered indigenous trees on Chimono and in the province of Mino. In this province, on the Nakasendō, below Nakatsu-gawa, we saw young plants of the Kôya-maki in all the roadside gardens, a pretty sure indication in this remote region that the tree was growing in the woods not very far off, and here for the next two or three days we saw it sending up its narrow pyramidal heads above the Pines and other trees of the forest, growing, as we thought, quite naturally, and leading us to believe that we had found the true home of this tree, although in a country like Japan, which has been densely populated for centuries, and in which tree-planting has been a recognized industry for more than a thousand years, it is not easy to determine whether a forest has been planted by man or not. But whether these trees had been planted or whether they were the offspring of trees brought from some other region, or the indigenous inhabitants of the forest, the Sciadopitys grows on the mountains of Mino in countless thousands, often rising with a tall straight trunk to the height of nearly a hundred feet, and remarkable in its narrow compact pyramidal head of dark and lustrous foliage. The wood, which is nearly white, strong, and straight-grained, is a regular article of commerce in this part of Japan, and from Nakatsu-gawa is floated in rafts down the Kisogawa to Osaka, where it is said to be chiefly consumed. Except in the neighborhood of Nakatsu-gawa, the Sciadopitys is not very much cultivated as a garden-plant in Japan; and it is not often found in old gardens, except in the immediate neighborhood of temples, where picturesque old specimens may occasionally be seen occupying a place of honor within the fence which incloses the principal buildings, and carefully protected by low stone railings. There is a remarkable specimen with pendulous branches standing before one of the mortuary temples in the Shiba Park in Tōkyō.

In Japan, Junipers are much less common than they are in eastern America, and although five or six species are included in the Floras of the empire, the genus does not make an important element of the landscape, and one misses the dark spires which Juniperus Virginiana sends up so frequently in many parts of eastern America. Juniperus Chinensis appears to grow to a larger size than the other Japanese species, although we saw it in only one region growing, as it appeared, without cultivation. This was on the high volcanic ridge which dominates the Chikuma, one of the streams which flow from Asama-yama, in central Japan. Over this elevated and inhospitable region occasional Junipers are scattered, the largest attaining a height of thirty or forty feet, their wind-swept heads and straggling branches, covered with gray-green foliage, adding to the dreariness of the scene. Before the Buddhist temple of Zenkōgi, in Nagano, the principal city in this part of Japan, two venerable Junipers show to what size plants of this species can grow, and how picturesque they can become. These trees are seventy or eighty feet high; their hollow trunks, which are rather more than six feet in diameter, support narrow heads of twisted and contorted branches clothed with scanty foliage, and indicate that centuries may well have passed since the roots of these marvelous trees first penetrated this sacred soil. On the rocky cliffs and grassy slopes of the coast, fully exposed to the spray of the ocean, a prostrate variety of Juniperus Chinensis forms, with its long creeping stems, dense mats, often of considerable size. It is said to be a feature of the littoral vegetation of Japan; but we saw only a few plants in Yezo, between Mororan and the Aino village of Horobetsu, where they receive the unbroken sweep of the Pacific. On the sandy dunes of the Bay of Hakodate, opposite that city, another littoral Juniper was found by the American botanist Charles Wright, and later by Maximowicz. This is the Juniperus conferta of Parlatore (Juniperus littoralis of Maximowicz), a species distinguished by its stout crowded leaves and large globose fruit. We saw it on Hakodate Bay at the end of September, and Mr. Veitch collected it earlier near Hanjo, on the west coast of Hondo, but we were too early to obtain ripe fruit of this species or of the prostrate form of Juniperus Chinensis. The only other Juniper we saw in Japan, Juniperus rigida, is a small tree sometimes twenty feet high, but more often a low spreading bush. It is common in the barrens near Gifu, and appears to be generally distributed at low elevations in central Japan, although it grows only on dry sterile gravelly soil. This is the Juniper which is most commonly cultivated by the Japanese, and which is not infrequently an inhabitant of temple gardens. Its long slender rigid leaves and small fruit, tipped with a minute mucro, serve to distinguish it from Juniperus conferta.

THE CONIFERS, II.

Of the true Pines of Japan two species are valuable timber-trees; these are Pinus densiflora and Pinus Thunbergii; both bear an important part in the decoration of Japanese gardens, and one at least has had its influence in all expressions of the artistic temperament of the people. All the Pine-woods of Japan, except those found on the upper slopes of some of the high mountains of central Hondo, have evidently been planted. Such planted woods are often seen covering sandy plains near the coast, and the principal highways of the empire are shaded by avenues of these Pines, except where Cryptomerias replace them when mountains are crossed. Of the two species the Black Pine, Pinus Thunbergii, appears to be the most commonly cultivated, and to grow to the larger size. Of its distribution and appearance when growing naturally I was not able to get an idea, as all the plants I saw had evidently been planted by man. It is of this species that the plantations of the coast are mostly formed, although the two species are generally found mixed together in all plantations; and it is this species which is usually selected by the Japanese gardener when he wants to make the branches of a Pine-tree cover an arbor or hang suspended over the sides of a moated wall, and which is found in every garden and is most revered by the Japanese. Pinus Thunbergii is one of the most picturesque of Pines, with a broad head of stout contorted somewhat pendulous branches, often growing to the height of eighty feet, and producing trunks three feet through. Its dark deeply furrowed bark, darker colored and thicker leaves and white buds serve to distinguish it from the Red Pine, Pinus densiflora, which is a tree of high altitudes, and which, although planted in large plantations and by the sides of highways, does not appear to be such a favorite in gardens as the Black Pine. The Red Pine we saw growing wild high up on Mount Koma-ga-take, in central Hondo, and on the Nikkō Mountains, where, at about 3,000 feet above the sea-level, it is not rare. It is a more slender tree than the Black Pine, with thinner lighter green leaves. The bark on the upper part of the trunk and on the main branches is light red, separating in thin scales, so that a forest of these trees presents a bright and cheerful appearance. Several varieties of the two species recognized by Japanese gardeners are described by Mayr, who also found what he thought was a hybrid between them.[1] The wood of the two species is very similar, and, apparently, is not distinguished in Japanese lumber-yards. It is coarse-grained, resinous, and moderately strong, and is used in great quantities in all sorts of coarse construction, and as fuel, the rapid growth of the trees on soil too poor to produce more valuable crops to advantage rendering it exceedingly cheap. These two Pines have long inhabited our gardens, where they are hardy and grow with great rapidity, some of the oldest plants of the Red Pine here already beginning to show the picturesque habit which in their native country constitutes the charm of these trees.

[1] *Die Abietineen des Japanischen Reiches*, 83, t. 7, f. 2, 3, 4.

The other Pines of Japan belong to the group in which the species produce their leaves in clusters of fives. The largest and the most widely distributed is Pinus parviflora, a beautiful small tree of high mountain forests, through which, at elevations over 5,000 feet above the sea-level, it is found scattered, either singly or in small groves, sometimes growing to a height of sixty or seventy feet, although it is usually much smaller. In those parts of Japan which we visited, it was most common and grew to the largest size on the slopes of Mount Hakkoda, in northern Hondo, where its dark pyramidal heads of slender spreading branches, rising above the forests of Oaks and Beeches, break the sky-line, just as its relative, our eastern White Pine, raises its nobler head high above the Oak forests of New England. The wood of Pinus parviflora is soft, straight-grained, light-colored, and of considerable value, but so difficult to obtain that it is little known or used by the Japanese. This beautiful Pine flourishes in our gardens, where it appears to be perfectly at home, and where it grows rapidly and every year covers itself with cones.

In southern Yezo, a second species of the same group, Pinus pentaphylla, has been distinguished by Mayr. This is an exceedingly rare tree, found in a few isolated situations, and distinguished from Pinus parviflora by its longer cones and stouter leaves. We saw only a cultivated tree at the hot springs of Kakumi, near the shore of Volcano Bay, being prevented by bad weather from reaching a small grove of these trees growing on the mountains in the neighborhood. This Pine has not been introduced into our gardens, where it may be expected to flourish.

The fifth Japanese Pine is interesting from the fact that it is the only Japanese Conifer which grows naturally in North America. It is the Pinus pumila of Regel, a species so similar to the Stone Pine of Europe that it has been considered by many authors a variety of that tree. We saw it only on the summit of Mount Hakkoda, where it forms, at 6,000 feet above the sea-level, impenetrable thickets a few feet in height and hundreds of acres in extent; it occurs on the summits of some of the high mountains of Yezo, ranges north through Saghalin and eastern Manchuria to Kamtschatka, and by the Kurile Islands reaches those of the Alaska coast.

Of Spruces, there appear to be four species in Japan, where, except, perhaps, in some parts of Yezo, they are exceedingly rare. The first, Picea polita, we saw only in two or three individuals in the Nikkō Mountains, on the hills below Lake Chuzenji. The trees were small, much torn and stunted by the wind, and of such a miserable appearance that it was difficult to realize that the young trees in perfect health and beauty which decorate our gardens belonged to the same species. For the second Spruce of the mountain forests of central Hondo, to which it appears to be confined, Mayr proposes the name of Picea bicolor, this specific name having, he finds, been first used by Maximowicz for this tree. This is the beautiful Spruce with blue-green leaves, silvery white on the under surface, which is usually cultivated under the name of Picea Alcockiana, and which is easily distinguished in the spring by the bright red color of the young shoots.

It is not my purpose to discuss here the synonymy of the Japanese Firs and Spruces, upon which such a mass of names have been heaped in almost hopeless confusion, that only a critical examination of all the specimens which have been studied by European botanists can

PLATE XXV.

HEMLOCK FOREST (TSUGA DIVERSIFOLIA, MAXM.), LAKE YUMOTŌ.

make it possible to reach any useful conclusions on the subject, and I shall only speak of the trees as I saw them growing in the forests of Japan. Picea bicolor, of which we saw but three or four specimens, is evidently a rare and local tree; found only at high elevations, scattered through the Oak and Beech forests, and, like Picea polita, presenting in its home a wretched and forlorn appearance. The leaves are nearly equally four-sided, and the cones are four to six inches long, with narrow pointed more or less laciniate scales. These two species, so far as I was able to observe, are the only Spruces which grow on the island of Hondo, the other species finding in Yezo their most southern home. These are Picea Ajanensis, a tree with smaller cones than the last, and short broad flat leaves, dark green above and pale on the lower surface. This is the common Spruce of Yezo, occurring on the hills near Sapporo, which is the only place where I saw it, in isolated individuals scattered through the forests of deciduous trees. According to Mayr, this tree forms in the western part of the island considerable forests on low swampy ground, not much raised above the level of the ocean. This appears to be the common Spruce of Saghalin and of the Manchurian coast.

The fourth species, Picea Glenhi, discovered by F. Schmidt in Saghalin, has been found in a few situations in southern Yezo. This tree, which is still to be introduced into our gardens, we did not see growing. In many characters it resembles the Siberian Picea obovata, and in the herbarium it is not easy to find characters by which it can be satisfactorily separated from that species. Like the species of the White Spruce group of North America, which appear to pass one into another by gradual transitions, the Spruces of northeastern Asia are difficult to distinguish with the material found in herbaria, and it will only be possible to study them satisfactorily when all the various forms have been planted side by side in some arboretum and allowed to grow to maturity.

Of the Hemlocks found in Japan, one is northern and the other southern; both are common at high elevations, and one at least forms extensive forests. The great forest, which covers the Nikkō Mountains at an altitude of more than 5,000 feet above the ocean, is composed almost entirely of the northern Hemlock, Tsuga diversifolia, which is distinguished by its bright red bark, small leaves, and cones. This Hemlock forest, which is the only forest in Hondo which seems to have been left practically undisturbed by man, is the most beautiful which we saw in Japan. The trees grow to a great size, and while they stand close together are less crowded than the trees in an American Hemlock forest, under which no other plants can grow, and light enough reaches the forest-floor to permit the growth of Ferns, Mosses, and many flowering under-shrubs which clothe the rocky slopes up which this forest stretches. One of the most beautiful spots which we saw in Japan is the walk cut through this forest which follows along the shores of Lake Yumoto, and I am fortunate in being able to reproduce a photograph of it (Plate xxv.), for which I am indebted to Professor Mayr, of Munich, who, during an official residence of several years in Japan, explored the forests of all parts of the empire more thoroughly than any other foreigner. We found Tsuga diversifolia in scattered groups on the rocky cliffs of Mount Hakkoda, in the extreme north of Hondo, the most northern station which has been recorded for this tree, which is still to be introduced into our gardens; but, south of Nikkō, it was replaced by the second species, Tsuga Tsuga, which we saw in great beauty on Koma-ga-take, where, however, it does not form a continuous forest, but is scattered in groves of considerable extent among deciduous trees and Pinus densiflora. It is this southern species, Tsuga Tsuga, which is cultivated in our gardens, where it appears to be as hardy as our native species, which it surpasses in its more graceful habit, and in its broader and darker colored leaves.

Of the Firs of Japan we saw only four species, and one of these, Abies firma, only as a cultivated plant. This is the largest and the most beautiful of the Japanese Firs, often growing in cultivation to the height of one hundred and twenty feet, and producing clean tall stems four or six feet in diameter. Writers on Japanese forests speak of Abies firma as common south of latitude 40°, north, in the upper belt of deciduous trees, but we never saw it except in parks and temple gardens, or in the immediate neighborhood of houses. It is this species which is chiefly called Momi by the Japanese, although the name is applied generally to all Firs, and which, in Hondo, supplies the fir wood of commerce. This is soft, straight-grained, and easily worked, and hardly distinguishable from the wood of the European Fir; it is used for building purposes and cheap packing-cases, but is not greatly valued. This species has usually proved a disappointment as an ornamental tree in this country and in Europe, but it is certainly, as it grows in Japan, one of the most beautiful of all Firs, distinguished by the nobility of its port and by its bright green and very lustrous long rigid leaves, which are sometimes sharply pointed, and sometimes divided at the apex. It probably needs a warmer and moister climate than that of the northern United States in which to develop all its beauties; farther south it should, however, make a fine tree.

The Fir of which we saw the most in Japan is the Abies homolepis of Siebold and Zuccarini. This is the plant which is now often cultivated in our gardens under the name of Abies brachyphylla, a more recent name. It is the common Fir of central Japan, and abounds on the Nikkō Mountains between 4,000 and 5,000 feet elevation above the sea, although it does not form continuous forests, but is scattered singly, or in small groups, through the Birch and Oak woods which cover the ground just below the Hemlock belt. It is a massive although not a very tall tree, apparently never growing to a greater height than eighty or ninety feet; in old age it is easily distinguished from all other Firs by its broad round head, the branches near the tops of the trees growing longer than those lower down on the stems. This peculiarity is seen even on young plants in our gardens, on which the lower branches, which soon stop growing, are shaded by the longer ones produced above them. The pale bark, the long crowded leaves, dark green above and silvery white below, and the large purple cones make this a handsome tree. In cultivation here it is very hardy, and grows with remarkable rapidity. The inaccessibility of the places where Abies homolepis grows in Japan precludes the general use of the wood, which we found employed in the little alpine village of Yumoto for building material.

The chief object of our visit to Mount Hakkoda, in northern Hondo, was to find Abies Mariesii, which the botanical collector, whose name this tree bears, discovered there several years before. It is common on this mountain at about 5,000 feet above the sea-level, scattered among deciduous trees, and, so far as we observed, it is the only Fir of northern Hondo. As we saw it, Abies Mariesii forms a compact pyramid about forty or fifty feet high, with crowded branches covered with short dark foliage, pale below, and many large dark purple cones. It is a handsome but in no wise a striking or remarkable tree, which in all probability will flourish in severe climates. It is only known on the high mountains of northern Hondo and in one place on the shores of southern Yezo, where it was discovered during the summer of 1892 by Mr. Tokubuchi.

On the hills of central Yezo, Abies Sachalinensis is not rare, and in the northern part of the island and on Saghalin this fine tree is said to form extensive forests. It is a tall pyramidal tree with pale bark, long slender dark green leaves, and white buds which make it possible to distinguish it readily from the other Japanese Firs. A curious form has been noticed by Professor Miyabe, growing near Sapporo, with red bark, dark red wood, and red cone bracts; it grows with the common form and is probably merely a seminal variety, although Professor Miyabe considers it specifically distinct and proposes to call it Abies Akatodo. We were fortunate in securing a supply of seeds of the white and of the red barked varieties, and the seedlings, perhaps, will show whether they should be considered distinct. Abies Sachalinensis produces wood of fair quality, which is used in Sapporo in building and for packing-cases; it has only a local consumption. The young plants of Abies Sachalinensis in our gardens are perfectly hardy and grow more rapidly than those of any other species of Fir-tree.

Abies Veitchii, discovered many years ago on the slopes of Mount Fuji-san, we looked for everywhere, and although it is said to grow among the Nikkō Mountains, we saw nothing of it there or elsewhere. From Abies homolepis, which this species most resembles, it may be distinguished by its shorter and more crowded leaves, its more slender pubescent shoots and smaller cones. This tree is an old inhabitant of our gardens, having been sent many years ago by Mr. Thomas Hogg to the Flushing Nurseries, where it was cultivated under the unpublished name of Abies Japonica long before it was known in Europe. Of Dr. Mayr's Abies umbellata, a species probably too near Abies homolepis, we saw nothing.

The forests of Hondo contain at least one Larch, Larix leptolepis. It is a fine tree, seventy or eighty feet tall, with pale green foliage and massive trunks covered with reddish bark, and in habit not unlike the European species. The Japanese Larch is not rare at elevations of from 5,000 to 6,000 feet in the central part of the island, although we saw it nowhere growing in continuous forests, but always scattered in small groves, mixed with other deciduous trees. Larix leptolepis was introduced into American gardens many years ago; it grows in this country with great rapidity, and the oldest trees here have for many years produced abundant crops of seeds. The wood, like that of other Larch-trees, is hard, heavy, and strong. The trees, however, are so difficult to reach that it is little used in Japan, except for the timbers of mountain mines.

Maximowicz describes a variety of this species, var. Murrayana (Larix Japonica, Murray), which grows as a low shrub near the timber-line of Fuji-san; and my companion, Mr. Codman, made a special trip late in the autumn for the purpose of securing specimens and seeds of this plant. In this he was successful, but his specimens gathered from plants but a few feet high, growing at an elevation of 8,500 feet above the sea-level, differ from those of the common arborescent form only in the smaller size of the cones and in the shorter leaves.

A variety of Larix Dahurica, a species widely distributed through Siberia, northern China, Manchuria, Kamtschatka, and Saghalin, reaches the extreme northern part of Yezo and the Kurile Islands. This form has been called var. Japonica by Maximowicz, and by Mayr Larix Kurilensis. We were not fortunate enough to see this tree in its native forests, but some idea of its appearance as it grows in the island of Iturup can be obtained from our illustration (Plate xxvi.), which is produced from a photograph for which I am indebted to Dr. Mayr, who visited the Kurile Islands during his residence in Japan.

The other arborescent plants of Japan, Cycas revoluta, a favorite garden-plant, especially in the south, where it often grows to a great size, and Trachycarpus (Chamærops) excelsa, naturalized in some parts of the south, appear to have been introduced from the Loochoo or other southern islands, or from Formosa or southern China.

PLATE XXVI.

LARIX DAHURICA, var. JAPONICA, Maxm.

THE ECONOMIC ASPECTS OF THE JAPANESE FORESTS.

In few other countries are the forests of greater importance to the prosperity of the nation than they are in Japan. The formation of the islands, with their high central mountain ranges and short precipitous swift flowing rivers, make floods particularly prevalent and dangerous, and the necessity of preserving the forest-covering of the upper mountain-slopes proportionately great; and no other race, with the single exception, perhaps, of the people of the United States, is such a consumer of forest products as the Japanese, all their houses and most of their articles of domestic use being entirely made of wood. The traveler, therefore, watches with some interest, as bearing upon the future of Japan, the condition and prospects of her forests.

According to the most reliable statistics available, those compiled by Rein, and based in part, at least, upon the report of the Japanese Forestry Exhibit at Edinburgh in 1884, thirty-seven per cent. of the three southern islands — that is, of the whole empire, with the exception of the practically unsettled island of Yezo — is desert or unproductive land. Twenty-three per cent. is occupied by the mountain forests, eighteen per cent. by the cultivated forests, while rather less than twenty per cent. is devoted to agriculture, the remainder being taken up by buildings, roads, etc. The cultivated forests, in which it is presumed that the areas surrounding the temples are included, which, although covered with splendid groves of trees, are unproductive except so far as they are made to furnish material for repairing or rebuilding the temples, are well stocked with coniferous trees, — Retinosporas, Cryptomerias, and Pines, — and furnish all the building material used in the empire. It is said that the Japanese have been making these plantations for twelve hundred years; and if this is true, they began planting trees for timber before any other people with whose agriculture we are acquainted. Scientific methods might, perhaps, make these plantations, which are mostly the property of individuals, rather more productive, but any great increase of forest supplies can only follow the better management of the mountain forests or the replanting of desert lands. The mountain forests, which are the property of the state, may be roughly divided into two belts, the upper composed principally of Hemlocks, and extending from about 5,000 feet above the sea-level to the timber-line, the lower stretching between 2,000 and 5,000 feet elevation, and composed of Beeches, Oaks, Maples, Birches, Pines, and a few Firs and Spruces. The upper belt, owing to its inaccessibility and the bad condition of the mountain roads, is practically untouched, except where mining operations have created a local demand for timber. Scientific management and good roads would make this upper coniferous forest yield quantities of valuable material. The deciduous belt below it, which ought to be the most productive part of the Japanese forest, is, wherever we entered it, in a deplorable condition, and although Japan has supplied herself with a Forest Department, we saw no evidence that it is seriously occupying itself with the care of this part of the public domain. There appear to be no rules about cutting the trees in this belt, which is invaded by bands of wood-choppers, who cut without system any trees that appear large enough to answer their purpose; and the only mature trees it contains are those growing in inaccessible positions, or of sorts which are not considered valuable. No attention is paid to reproducing the valuable species, with the result, of course, that such species, being the most cut, have become the least common. Reproduction is chiefly by coppice-growth which is cut at irregular intervals; and the Japanese deciduous forests display all the bad effects of an indiscriminate and long-continued system of jardinage. The application of the block system would in time, of course, increase the output of these forests and supply large quantities of valuable timber where only fuel or small sticks are now produced.

Still better results might be expected from covering some parts of the desert land with forests. These so-called deserts consist of sandy seashore planes and dunes, often capable of producing a moderate growth of Pine; the alpine summits of mountains, their lava-covered slopes, bare mountain-ridges from which the forests have been artificially removed, and the hara. This is the rolling foothill region about the base of the high mountains or below the mountain forest-belt, and must form a very large part of the thirty-seven per cent. of desert; it is covered with a mat of coarse bunch-grass (Eulalia) and with many other perennial plants. Here the Japanese cut the fodder for their animals and cure their hay for winter use. Every spring the whole hara is burnt over to destroy the dried vegetation of the previous year and start a new growth of grass. That fires have made these foothills treeless by the destruction of all seedling trees as fast as they appear seems to be shown in the fact that where ravines or other depressions occur among the hills, which the fire cannot easily reach, they are covered with a vigorous growth of trees of many species. It is not easy to find in existing conditions of Japanese life any cause for the original destruction of the foothill forests, but once destroyed it is easy to see why they have not been able to grow again. Much of the hara region is suited in soil and elevation to produce Retinosporas and Cryptomerias, the most valuable of the Japanese timber-trees, and its conversion from unproductive prairie — for not one per cent. of the hara is used for hay — into forest would add enormously to the wood product of the empire.

Japan is well situated geographically to supply a vast number of people living in foreign countries with timber; and its soil and climate are preëminently suited to produce forests. It could easily send, if it had it to spare, coniferous timber to China, where the demand for building material is practically inexhaustible, to the Straits Settlements and Australia; and oak-staves for wine casks to California, which is now supplied from the fast vanishing forests of the Mississippi valley.

From the changed conditions which have followed the hasty and often ill-considered introduction of European methods into Japan, grave economic questions are rising. The cessation of civil wars which followed the abolition of the Shogunate and the deposition of the Daimyos, and the introduction of western medical practices, have caused a great increase of population in the last twenty years, and the question of food supply is becoming a vital one to Japan. The limit to the production of rice, the one great staple, has been practically reached, and all efforts to induce the superfluous population of the southern islands to colonize Yezo have utterly failed, in spite of the great sums of money spent by the Imperial government during the last twenty-five years to encourage its settlement. A few thousand coolies leave home annually to work in other countries, but this movement is comparatively small, and many of these emigrants return to their homes at the end of a few years. Starvation threatens Japan unless it can import food from other countries, and this it will only be able to do by increasing its exports. There is still room to increase the product of tea if the demand in this country for low grades of Japanese tea justifies it, but the ground fit to grow Mulberry-trees advantageously is practically all taken up, and the silk product cannot therefore be very materially increased. Curios, of course, can be made in unlimited quantities, but the demand for them in the United States and Europe is more likely to decrease than to increase; and wood is really the only product upon which Japan can depend to greatly increase the volume of her exports. The care of her existing forests and the planting of her waste lands would give employment to thousands of coolies, and in time would add important sums to the national exchequer.

The forests of Yezo are still intact, except where here and there a struggling settlement has broken into the forest-blanket which covers this noble island. Here are great supplies of oak and ash of the best quality, of cercidiphyllum, walnut, fir, acanthopanax, cherry and birch — a storehouse of forest wealth, which, if properly managed, could be drawn upon for all time, and which, if the timber is not needed in Japan, may become, when the trans-Asiatic railroad is finished, an important factor in the development of southern Siberia and some of the treeless countries of central Asia.

INDEX.

Names of admitted Genera and Species and other Proper Names in roman type; of Synonyms in *italics*.

Abies Alcoquii, 83.
Abies brachyphylla, 82.
Abies, distribution of, in Japan, 6.
Abies firma, 82.
Abies firma in American gardens, 82.
Abies firma, wood of, 82.
Abies homolepis, 82.
Abies homolepis, distribution of, 82.
Abies Japonica, 83.
Abies Mariesii, 10, 82.
Abies Sachalinensis, 83.
Abies Sachaliuensis, red wooded variety of, 83.
Abies umbellata, 83.
Abies Veitchii, 83.
Acanthopanax, 44.
Acanthopanax aculeatum, 44.
Acanthopanax innovans, 44.
Acanthopanax ricinifolium, 44, 45.
Acanthopanax ricinifolium, wood of, 46.
Acanthopanax sciadophylloides, 44.
Acanthopanax sessiliflorum, 44.
Acanthopanax trichodon, 44.
Acer argutum, 31.
Acer capillipes, 30.
Acer carpinifolium, 30.
Acer cissifolium in American gardens, 31.
Acer crataegifolium, 31.
Acer diabolicum, 31.
Acer distylum, 31.
Acer Japonicum, 30.
Acer Japonicum, foliage of, in autumn, 30.
Acer Maximowiczianum, 32.
Acer Miyabei, 29.
Acer Nikoense, 31.
Acer Nikoense in European gardens, 32.
Acer palmatum, 30.
Acer palmatum, varieties of, in Japanese gardens, 30.
Acer parvifolium, 31.
Acer pictum, 28.
Acer purpurascens, 31.
Acer pycnanthum, 31.
Acer rufinerve, 31.
Acer Sieboldianum, 31.
Acer spicatum, 30.
Acer Tataricum, var. Ginnala, 30.
Acer Tataricum, var. Ginnala, in American gardens, 30.

Acer trifidum, 1.
Acer Tschonoskii, 31.
Actinidia arguta, 7, 19.
Actinidia arguta, fruit of, 19.
Actinidia in Japan, 12.
Actinidia Kolomikta, 19.
Actinidia polygama, 19.
Actinidia polygama, peculiarity of the leaves of, 19.
Æsculus in North America, 28.
Æsculus turbinata, 28.
Albizzia Julibrissin, 34.
Alder, the black, 25.
Alders, arborescent, abundance of, in Japan, 7.
Alders in Japan, 63.
Aleurites cordata, 56.
Alnus firma, 63
Alnus firma, 63.
Alnus firma, var. multinervis, 63.
Alnus incana, 63.
Alnus incana, var. glauca, 63.
Alnus incana, var. hirsuta, 63.
Alnus Japonica, 63.
Alnus maritima, 63.
Alnus viridis, var. Sibirica, 63.
American forests, composition of, compared with Japanese, 1.
Andromeda Japonica, 49.
Andromeda nana, 10.
Aphananthe aspera, 59.
Aralia Chinensis, 44.
Aralia cordata, 44.
Aralia family in Japan, 43.
Aralia family in eastern America, 43.
Aralia family in North America, 43.
Aralia Maximowiczii, 45.
Aralia quinquefolia, 43.
Aralia racemosa, 44.
Aralia spinosa, var. elata, 44.
Arborescent genera, common to the forests of eastern North America and Japan, 6.
Arbor vitæ, the Japanese, 72.
Aria in Japan, 39.
Aronia alnifolia, 40.
Ash, the Mountain, 78.
Aspen, the, 71.
Azalea calendulacea, 49.
Azalea mollis, 49.
Azaleas in Japan, 49.

Balsam Poplar, the, in Japan, 71.

Bamboos, prevalence of, in Japanese forests, 7.
Beech, the, in Japan, 70.
Beech-wood in Japan, 70.
Benthamia, 47.
Berberis Japonica, 2.
Berchemia racemosa, 26.
Betula alba, 61.
Betula alba, var. Tauschii, 62.
Betula alba, var. verrucosa, 62.
Betula Bhojpattra, 63.
Betula corylifolia, 63.
Betula Ermani, 7, 62.
Betula Maximowicziana, 62.
Betula serra, 63.
Betula ulmifolia, 63.
Birches, abundance of, in Japan, 7.
Birch-trees, black-barked, 62.
Birch-trees in Japan, 61.
Black Alder, 25.
Black Pine, 79.
Broussonetias in Japan, 59.
Bunch-berry, the, 47.
Bürgeria? salicifolia, 11.

Camellia Japonica, 17.
Camellia Japonica, oil of, 17.
Camellia Japonica, wood of, 17.
Camellia Sasanqua, 17.
Camellia theifera, 17.
Camphor-tree, the, 54.
Camphor-tree at Atami, 54.
Carpinus Carpinus, 64.
Carpinus Carpinus in American gardens, 65.
Carpinus cordata, 65.
Carpinus erosa, 64.
Carpinus Japonica, 64.
Carpinus laxiflora, 64.
Carpinus Tschonoskii, 64.
Carpinus Yedoensis, 64.
Carpinus, distribution of, 64.
Catalpa ovata, 2.
Cedrela Sinensis, 1.
Celtis Sinensis, 59.
Cephalotaxus drupacea, 75.
Cercidiphyllum Japonicum, 13.
Cercidiphyllum from western China, 4.
Cercidiphyllum, its introduction into American gardens, 15.
Cercidiphyllum, wood of, 15.
Cercis Chinensis, 2.

INDEX.

Celastrus articulatus, 26.
Celastrus articulatus, decorative uses of, in Japan, 26.
Celastrus flagellaris, 26.
Chamæcyparis obtusa, 73.
Chamæcyparis obtusa, plantations of, in Japan, 73.
Chamæcyparis obtusa, uses of the wood and bark of, in Japan, 73.
Chamæcyparis obtusa in the eastern United States, 73.
Chamæcyparis pisifera, 73.
Chamæcyparis pisifera in the eastern United States, 73.
Chamæcyparis pisifera, wood of, 73.
Chamærops excelsa, 3, 84.
Cherry, the Japanese, 36.
Cherry-tree cultivated in Japan, 36.
Cherry-tree, Pendulous-branched, 37.
Cherry-tree, varieties cultivated in Japan, 36.
Chestnut-tree, the, in Japan, 60.
Chestnuts as food in Japan, 70.
Chinese rice-paper, 43.
Chinese shrubs cultivated in Japan, 2.
Chinese trees cultivated in Japan, 1.
Chionanthus, 52.
Chionanthus retusa, 2.
Cinnamomum Camphora, 54.
Cinnamomum pedunculatum, 54.
Citrus Japonica, 2.
Clematis patens, 2.
Clerodendron trichotomum, 53.
Clethra barbinervis, 50.
Clethra canescens, 50.
Cleyera ochnacea, 17.
Climbing plants in Japanese forests, 7.
Climbing Hydrangeas in Japan, 7.
Composition of Japanese and eastern American forests compared, 1.
Conifers in Japan, 72.
Conifers planted in Japan, 72.
Corean shrubs cultivated in Japan, 2.
Corean trees cultivated in Japan, 1.
Cornus alba, 47.
Cornus brachypoda, 48.
Cornus Canadensis, 7, 47.
Cornus Kousa, 47.
Cornus Kousa in American gardens, 47.
Cornus macrophylla, 47, 48.
Cornus officinalis, 1, 48.
Cornus officinalis cultivated in Japan, 48.
Cornus Suecica, 47.
Cornus in North America, 47.
Cratægus alnifolia, 40.
Cratægus chlorosarca, 41.
Cratægus cuneata, 1.
Cratægus in eastern North America, 41.
Cryptomeria Japonica, 74.
Cryptomeria Japonica, plantations of, in Japan, 74.
Cryptomeria Japonica, groves and avenues of, at Nikkō, 74.
Cryptomerias at Nikkō, age of, 75.
Cryptomeria Japonica, uses of the wood and bark of, in Japan, 74.

Cupuliferæ in Japan, 61.
Cycas revoluta, 3, 84.

Daphne Genkwa, 2.
Daphniphyllum glaucescens, 56.
Daphniphyllum humile, 10, 56.
Daphniphyllum macropodum, 56.
Dendropanax Japonicum, 43.
Desert lands, areas of, in Japan, 85.
Desert lands, nature of, in Japan, 86.
Diervilla Japonica, 48.
Dimorphanthus Manchuricus, 44.
Diospyros Kaki, 2, 50.
Diospyros Lotus, 2, 51.
Disanthus cercidifolia, 42.
Distigocarpus, 64.
Distigocarpus Carpinus, 64.
Distylium racemosum, 42.

Economic aspects of Japanese forests, 85.
Edgeworthia papyrifera, 2.
Ehretia acuminata, 53.
Elæagnus longipes, 56.
Elæagnus pungens, 56.
Elæagnus umbellata, 56.
Elæagnus umbellata in American gardens, 56.
Elæocarpus in Japan, 21.
Elæocarpus photiniæfolia, 21.
Eleutherococcus in Japan, 43.
Elliottia paniculata, 7.
Elm-bark cloth, made in Yezo, 57.
Elm-family in Japan, 57.
English walnuts, 60.
Enkianthus campanulatus, 49.
Enkianthus Japonicus, 2.
Epigæa Asiatica, 10.
Ericaceæ in Japan, 49.
Eriobotrya Japonica, 1.
Euphorbiaceæ in Japan, 56.
Euptelea polyandra, 15.
Eurya Japonica, 17.
Evergreen shrubs, prevalence of, in Japan, 6.
Evodia rutæcarpa, 21.
Evodia rutæcarpa, economic uses of, 21.
Evonymus alatus, 26.
Evonymus alatus in American and Japanese gardens, 26.
Evonymus alatus, var. subtriflora, 26.
Evonymus Europæus, var. Hamiltonius, 26.
Evonymus Europæus, var. Hamiltonianus in American gardens, 26.
Evonymus in Japan, 25.
Evonymus Japonicus, 25.
Evonymus macropterus, 26.
Evonymus Nipponicus, 26.
Evonymus oxyphyllus, 26.
Evonymus radicans, 25.
Evonymus, the climbing, 7.
Excœcaria Japonica, 57.

Fagus Sieboldii, 70.
Fagus sylvatica, 70.
Fagus sylvatica, var. Sieboldii, 70.

Fatsia horrida, 43.
Fatsia Japonica, 43.
Fatsia papyrifera, 43.
Features of vegetation in Japan and eastern North America compared, 4.
Fig-trees in Japan, 59.
Firs in Japan, 82.
Forestiera, 52.
Forest-region, the Atlantic, extent of, 2.
Forest-region, the Japanese-Manchurian, 2.
Forests, Japanese, character and divisions of, 85.
Forests, Japanese, composition of, compared with eastern American, 1.
Forests, Japanese, economic aspects of, 85.
Forests, Japanese, methods of reproduction, 86.
Forests of Yezo, 87.
Forest, the Atlantic, census of, 2.
Forest, the Hemlock, of Lake Yumoto, 7.
Forest, the Japanese-Manchurian, census of, 2.
Forest undergrowth, character of, in Japan, 7.
Forest undergrowth, character of, in North America, 7.
Forsythia suspensa, 2.
Fraxinus in eastern America, 52.
Fraxinus longicuspis, 52.
Fraxinus Manchurica, 52.
Fraxinus Manchurica in American gardens, 52.
Fraxinus pubinervis, 52.

Gaultheria pyroloides, 10.
Genera, arborescent, common to the forests of eastern North America and Japan, 6.
Genera, endemic in eastern North America, list of, 4.
Genera, endemic in Japan, list of, 4.
Genera, number of arborescent, in eastern North America, 4.
Genera, number of arborescent, in Japan, 4.
Geum dryadoides, 10.
Ginkgo biloba, 2.
Gingko, fruit of, 76.
Gingko, the, 75.
Gingko, the, in temple grounds of Tōkyō, 76.
Ginseng, cultivation of, in Japan and Corea, 43.
Ginseng, the, 43.
Gleditsia Japonica, 35.
Gleditsia Japonica, use of the fruit in Japan, 35.
Grape, the wild, in the forests of Yezo, 7.

Ilnkkoda, Mount, vegetation of, 10.
Hamamelis arborescens, 42.
Hamamelis Japonica, 42.
Hamamelis, the Japanese, 42.
Hara, the, 86.

INDEX. 91

Helwingia in Japan, 43.
Hemlock forest of Hondo, 81.
Hemlock forest of Lake Yumoto, plants in, 7.
Hemlock, forests of, in Japan, 6.
Hemlocks, distribution of in Japan, 81.
Hi-no-ki, the, 73.
Hollies, the, in Japan and North America, 23.
Honeysuckle family, the, in Japan, 48.
Hop Hornbeam, distribution of, 65.
Horse-chestnut, the Japanese, 28.
Horse-chestnut, use of the fruit in Japan, 28.
Hovenia dulcis, 26.
Hovenia dulcis, cultivated as a fruit-tree in Japan, 26.
Hovenia dulcis, cultivated in America and Europe, 27.
Hydrangea paniculata, 41.
Hydrangea petiolaris, 7.
Hydrangeas, climbing, in Japan, 7.

Ilex crenata, 10, 24.
Ilex crenata in Japanese gardens, 24.
Ilex geniculata, 25.
Ilex integra, 23.
Ilex integra, var. leucoclada, 10, 23.
Ilex latifolia, 23.
Ilex macropoda, 25.
Ilex Monticola, 25.
Ilex pedunculosa, 24.
Ilex pedunculosa in Japanese gardens, 24.
Ilex rotunda, 20.
Ilex serrata, 25.
Ilex Sieboldii, 25.
Ilex Sieboldii in American gardens, 25.
Ilex Sieboldii, used for decoration in Japan, 25.
Ilex Sugeroki, 10, 24.
Ilex verticillata, 25.
Illicium anisatum, 12.
Illicium religiosum, 12.
Illicium, the Japanese, in religious festivals, 12.
Illicium verum, 12.
Ivy, the, in Japan, 7, 43.
Ivy, the, Poison, 34.

Japan as a producer of timber, 85.
Japanese Arbor-vitæ, 72.
Japanese Cherry, 36.
Japanese and North American forests, composition of, compared, 1.
Japanese Lacquer-tree, 33.
Japanese Plums cultivated in the United States, 38.
Juglans cordiformis, 60.
Juglans regia, cultivated in Japan, 60.
Juglans Sieboldiana, 60.
Juglans Sieboldiana in American gardens, 60.
Junipers in Japan, 78.
Juniperus Chinensis, 78.
Juniperus Chinensis at Nagano, 78.

Juniperus Chinensis, prostrate form of, 78.
Juniperus conferta, 78.
Juniperus littoralis, 78.
Juniperus rigida, 78.

Kadsura Japonica, 12.
Kaki, the, 50.
Kaki, varieties of, 50.
Kaya-no-abura, 76.
Kaya, the, 76.
Keaki, the, 57.
Kobus, 9.
Kœlreuteria paniculata, 1.
Kōya-maki, 77.

Lacquer-tree, Japanese, 33.
Larix Dahurica, distribution of, 84.
Larix Dahurica, var. Japonica, 84.
Larix Japonica, 83.
Larix Kurilensis, 84.
Larix leptolepsis, 83.
Larix leptolepsis in American gardens, 83.
Larix leptolepsis, var. Murrayana, 83.
Larix leptolepsis, wood of, 83.
Lauraceæ in Japan, 54.
Leguminosæ in Japan, 34.
Leitneria, 4.
Lignstrum Ibota, 53.
Lignstrum Japonicum, 53.
Lignstrum medium, 53.
Lindera Benzoin, 54.
Lindera glauca, 56.
Lindera in eastern Asia, 54.
Lindera, melissæfolia, 54.
Lindera obtusiloba, 55.
Lindera præcox, 55.
Lindera sericea, 55.
Lindera strychnifolia, 2.
Lindera triloba, 55.
Lindera umbellata, 55.
Liquidambar Formosana, 1.
Liquidambar Maximowiczii, 1.
Litsea glauca, 14.

Maackia Amurensis, 34.
Maackia Amurensis, wood of, 34.
Machilus Thunbergii, 54.
Magnolia compressa, 11.
Magnolia conspicua, 1.
Magnolia family, the, 8.
Magnolia family, the, in Japan and Manchuria, 8.
Magnolia family, the, in North America, 8.
Magnolia fuscata, 11.
Magnolia glauca, var. *a*, 9.
Magnolia hypoleuca, 8.
Magnolia hypoleuca as a timber-tree, 9.
Magnolia hypoleuca, introduction of, into American gardens, 9.
Magnolia Kobus, 9.
Magnolia Kobus, introduction of, into American gardens, 10.
Magnolia obovata, 2.

Magnolia parvifolia, 1.
Magnolia salicifolia, 10.
Magnolia stellata, 2.
Magnolia Thurberi, 10.
Magnolia tomentosa, 9.
Magnolia Watsoni, 1.
Maples in Japan, 28.
Maples, Japanese, in American gardens, 30.
Marrons of Kobe, 70.
Meliosma, 33.
Meliosma myriantha, 33.
Michelia compressa, 11.
Michelia, distinctive characters of, 11.
Michelia fuscata, 11.
Mohrodendron Carolinum, 51.
Momi, the, 82.
Monkey-slide, the, 18.
Morus alba, 59.
Mountain Ash, 38.
Mulberry, the White, 59.
Myrica Gale, 61.
Myrica rubra, 61.

Nandina domestica, 2.
Nara, grove of Podocarpus Nageia, at, 77.
Negundo in Japan, 31.
Nikkō, groves and avenues of Cryptomeria Japonica at, 74.

Oaks, evergreen, in Japan, 67.
Olea fragrans, 2.
Olive family, the, in Japan, 52.
Oriza Japonica, 21.
Osmanthus Aquifolium, 52.
Osmanthus Aquifolium in Japanese gardens, 52.
Osmanthus ilicifolium, 52.
Ostrya Japonica, 66.
Ostrya, the distribution of, 66.
Ostrya Virginica, var. *Japonica*, 66.

Panax horrida, 7.
Panax repens, 43.
Panax ricinifolia, 45.
Paper Mulberries in Japan 59.
Paulownia imperialis, 2.
Pear-tree of Japan, 40.
Pendulous branched Cherry-tree, 37.
Persimmon, the, in Japan, 20.
Phellodendron Amurense, 21.
Phyllodoce taxifolia, 10.
Picea Ajanensis, 81.
Picea Ajanensis, distribution of, 81.
Picea Alcockiana, 80.
Picea, bicolor, 80, 81.
Picea, distribution of in Japan, 6.
Picea Glenhi, 81.
Picea polita, 80.
Picrasma quassioides, 22.
Pines, Japanese, in American gardens, 79.
Pine, the Black, 79.
Pine, the Red, 79.
Pine, the Umbrella, 77.
Pine-wood, nature and uses of, in Japan, 79.

92 INDEX.

Pine-woods, planted in Japan, 70.
Pinus densiflora, 79.
Pinus, distribution of, in Japan, 72.
Pinus Koraiensis, 2.
Pinus parviflora, 80.
Pinus parviflora in American gardens, 80.
Pinus pentaphylla, 80.
Pinus pumila, 80.
Pinus pumila on Mount Hakkoda, 10.
Pinus, scarcity of species in Japan, 6.
Pinus Thunbergii, 79.
Plantations connected with Buddhist temples, 5.
Plants, climbing, in Japanese forests, 7.
Platycarya strobilacea, 41.
Plums, Japanese, cultivated in the United States, 38.
Plum-trees in eastern North America, 38.
Plum-trees in Japan, 38.
Podocarpus, distribution of, 77.
Podocarpus macrophylla, 2, 77.
Podocarpus Nageia, 2, 77.
Podocarpus Nageia, variegated form in Japanese gardens, 77.
Poison Ivy, the, 34.
Poplar, the Balsam, in Japan, 71.
Populus in Japan, 71.
Populus Sieboldii, 61.
Populus suaveolens, 71.
Populus tremula, var. villosa, 71.
Prunus Grayana, 38.
Prunus Japonica, 2.
Prunus Maximowiczii, 37.
Prunus Mume, 1, 36.
Prunus Mume in Corea, 36.
Prunus Padus, 38.
Prunus Pseudo-Cerasus, 36.
Prunus Pseudo-Cerasus, wood of, 37.
Prunus Ssiori, 38.
Prunus Ssiori, uses of the wood of, 38.
Prunus tomentosa, 2.
Prunus triflora, 38.
Pterocarya, distribution of, 61.
Pterocarya rhoifolia, 61.
Pterocarya rhoifolia, wood of, 61.
Pterostyrax corymbosum, 51.
Pterostyrax hispidum, 51.
Pyrus alnifolia, 57.
Pyrus aucuparia, 38.
Pyrus gracilis, 39.
Pyrus Halleana, 40.
Pyrus lanata, 39.
Pyrus Malus floribunda, 40.
Pyrus microcarpa, 40.
Pyrus Miyabei, 40.
Pyrus Parkmani, 40.
Pyrus Ringo, 40.
Pyrus sambucifolia, 38.
Pyrus Sieboldii, 40.
Pyrus Sinensis, 1, 40.
Pyrus Toringo, 40.
Pyrus Toringo in American gardens, 40.
Pyrus Tschonoskii, 40.

Quercus acuta, 69.
Quercus crispula, 67.
Quercus cuspidata, 69.
Quercus Daimio, 67.
Quercus dentata, 67.
Quercus dentata, var. pinnatifida, 67.
Quercus gilva, 69.
Quercus glabra, 69.
Quercus glandulifera, 68.
Quercus glandulifera in American gardens, 68.
Quercus glauca, 69.
Quercus grosseserrata, 67.
Quercus in Japan, 67.
Quercus lacera, 69.
Quercus serrata, 68.
Quercus Thalassica, 69.
Quercus variabilis, 68.
Quercus Vibrayana, 69.

Red Pine, the, 79.
Retinospora obtusa, 73.
Retinosporas, monstrous forms of, in Japan, 74.
Rhamnaceæ, 26.
Rhododendron arborescens, 49.
Rhododendron Catawbiense, 7, 49.
Rhododendrons, evergreen in Japan, 49.
Rhododendron maximum, 49.
Rhododendron Metternichii, 7.
Rhododendron mudicaule, 49.
Rhododendron Sinense, 49.
Rhododendron viscosum, 49.
Rhodotypos kerreoides, 2.
Rhus family, the, in eastern North America, 33.
Rhus family, in Japan, 33.
Rhus semialata, 33.
Rhus succedanea, 34.
Rhus sylvestris, 34.
Rhus trichocarpa, 24.
Rhus vernicifera, 1, 33.
Rice-paper, Chinese, 43.
Rose family, the, in Japan, 36.
Rue family, the, 21.

Sabiaceæ, 33.
Salix Babylonica, 70.
Salix ericocarpa, 71.
Salix subfragilis, 71.
Sapindaceæ, 28.
Sapindus Mukorosi, 1.
Saru-suberi, 18.
Sasan-kuwa, 17.
Sawara, the, 73.
Saxifrage family in Japan, 41.
Schizandra, 8.
Schizandra Chinensis, 13.
Schizandra nigra, 13.
Schizophragma, 7.
Sciadopitys cultivated in Japan, 77.
Sciadopitys, the, 77.
Sciadopitys, the wood of, 77.
Shrubs, Chinese, cultivated in Japan, 2.

Shrubs, Corean, cultivated in Japan, 2.
Shrubs, evergreen, in northern Japan, 23.
Shrubs, evergreen, prevalence of, in Japan, 6.
Shrubs, number of, in Japan, 3.
Simarubæ, 22.
Skimmia Japonica, 21.
Sophora Japonica, 1.
Sorbus alnifolia, 40.
Sorbus Aria, var. *Kamaonensis*, 39.
Sorbus lanata, 39.
Spice-bush, the, 54.
Spikenard, the North American, 44.
Spiræa Thunbergii, 2.
Spruces in Japan, 80.
Stachyurus præcox, 18.
Stachyurus præcox in American gardens, 18.
Sterculia platinifolia, 1.
Stuartia in America, 18.
Stuartia monadelpha, 18.
Stuartia pentagyna, 18.
Stuartia Pseudo-Camellia, 18.
Stuartia Pseudo-Camellia in American gardens, 18.
Stuartia serrata, 18.
Stuartia Virginica, 18.
Styraceæ in Japan, 51.
Styrax Japonica, 52.
Styrax Obassia, 51.
Sugi, the, 74.
Syringa Japonica, 52.
Syringa Japonica in New England gardens, 52.

Taxus cuspidata, 76.
Taxus cuspidata in American gardens, 76.
Taxus cuspidata, wood of, 76.
Taxus, distribution of, 76.
Tea-plant, the, 17.
Tecoma grandiflora, 2.
Temples, Buddhist, plantations connected with, 5.
Ternstræmiaceæ, 17.
Ternstræmia Japonica, 17.
Thuja Japonica, 72.
Thuja orientalis, 2.
Thuyopsis dolobrata, 72.
Thuyopsis dolobrata, varieties of, 72.
Thuyopsis Standishii, 72.
Tilia cordata, var. *Japonica*, 20.
Tilia heterophylla, 20.
Tilia Mandshurica, 20.
Tilia Miqueliana, 19.
Tilia ulmifolia, var. Japonica, 20.
Tilia ulmifolia, var. Japonica, in American gardens, 20.
Torreya, 76.
Toxylon, 4.
Trachycarpus excelsa, 3, 84.
Treeless foothill region, 86.
Trees, aggregation of species, in southern Indiana, 3.
Trees, aggregation of species in Japan, 3.

INDEX.

Trees, American, list of, absent from the forests of Japan, 5.
Trees, Chinese, cultivated in Japan, 1.
Trees, Corean, cultivated in Japan, 1.
Trees, Japanese, list of, absent from the forests of eastern North America, 6.
Trees, list of, on hills near Sapporo, 3.
Trochodendraceæ, 13.
Trochodendron aralioides, 15.
Tsubaki, 17.
Tsuga diversifolia, 7, 81.
Tsuga diversifolia, northern station of, 10.
Tsuga Tsuga, 81.
Tsuga Tsuga in American gardens, 82.
Tumion, distribution of, 76.
Tumion nuciferum, 76.
Tumion nuciferum, fruit of, 76.
Tumion nuciferum, wood of, 76.
Types of vegetation common to the floras of eastern North America and Japan, 5.

Ulmus campestris, 57.
Ulmus parvifolia, 2.
Ulmus scabra, laciniata, 17.
Umbrella Pine, the, 77.
Undergrowth of the forests of eastern North America and Japan contrasted, 7.

Vaccinium ciliatum, 49.
Vaccinium in Japan, 49.
Vaccinium Japonicum, 49.
Vegetation, features of, in eastern North America and Japan compared, 4.
Vegetation of Mount Hakkoda, 10.
Vegetation, types of, common to the floras of eastern North America and Japan, 5.
Viburnum furcatum, 48.
Viburnum Wrightii, 49.
Vitis Coignetiæ, 46.

Walnuts as food in Japan, 60.
Walnuts, English, 60.
Walnut family, the, in Japan, 60.

White Birch, the, 61.
Wikstrœmia Japonica, 2.
Willows, arborescent, abundance of, in Japan, 7.
Willows in Japan, 70.
Willow, the Weeping, planted in Tōkyō, 70.
Wistaria in Japan, 7.
Witch Hazel family in Japan, 41.

Xanthoxylum ailanthoides, 21.
Xanthoxylum in Japan, 21.
Xanthoxylum piperitum, 21.
Xanthoxylum piperitum, uses of the fruit in Japan, 21.

Yezo, forests of, 87.

Zelkova, distribution of, 58.
Zelkova Keaki, 57.
Zelkova Keaki in Rhode Island, 58.
Zelkova Keaki, wood of, 58.
Zizyphus vulgaris, 1.
Zizyphus vulgaris, cultivated as a fruit-tree in Japan, 2.

www.ingramcontent.com/pod-product-compliance
Lightning Source LLC
Chambersburg PA
CBHW030344170426
43202CB00010B/1235